Current and Future Aspects of Nanomedicine

Edited by Islam Ahmed Hamed Khalil

Published in London, United Kingdom

IntechOpen

Supporting open minds since 2005

Current and Future Aspects of Nanomedicine
http://dx.doi.org/10.5772/intechopen.83043
Edited by Islam Ahmed Hamed Khalil

Contributors
Mark D. Scott, Kerryn Matthews, Hongshen Ma, Mohd Ridzuan Ahmad, Muhammad Asraf Mansor, Mahmoud El-Badry, Khaled Abulftooh, Ayat Allam, Lucie Bacakova, Marketa Zikmundova, Julia Pajorova, Elena Filova, Petr Mikes, Vera Jencova, Eva Kuzelova Kostakova, Alla Sinica, Mesut Karahan, Shen Leng Tan, Öznur Özge Özcan, Palanirajan Vijayaraj Kumar, Yi Na Tee, Philip Serwer, Elena Wright, Cara Gonzales, Islam Khalil

First published in London, United Kingdom, 2020 by IntechOpen
IntechOpen is the global imprint of INTECHOPEN LIMITED, registered in England and Wales, registration number: 11086078, 7th floor, 10 Lower Thames Street, London, EC3R 6AF, United Kingdom
Printed in Croatia

British Library Cataloguing-in-Publication Data
A catalogue record for this book is available from the British Library

Additional hard and PDF copies can be obtained from orders@intechopen.com

Current and Future Aspects of Nanomedicine
Edited by Islam Ahmed Hamed Khalil
p. cm.
Print ISBN 978-1-78985-869-3
Online ISBN 978-1-78985-870-9
eBook (PDF) ISBN 978-1-78984-438-2

For EU product safety concerns:
IN TECH d.o.o., Prolaz Marije Krucifikse Kozulić 3, 51000 Rijeka, Croatia, info@intechopen.com or visit our website at intechopen.com.

Meet the editor

Islam Ahmed Hamed Khalil is an associate professor of Pharmaceutics at the College of Pharmacy and Drug Manufacturing, Misr University for Science and Technology, Egypt. He received a BS in Pharmaceutical Sciences from Misr University for Science and Technology in 2003, and a MSc (2009) and PhD (2013) in Pharmaceutics from the Faculty of Pharmacy, Cairo University. He worked as a postdoctoral research fellow at the American University in Cairo (Egypt), Zewail City for Science and Technology (Egypt), Brigham and Women's Hospital of Harvard Medical School (USA), and Northeastern University (USA). Dr. Khalil has also worked as a consultant in the pharmaceutical industry and bioequivalent centers. His research focuses on designing, developing and evaluating nanomedicines for various biomedical applications such as therapeutics and tissue regeneration.

Contents

Preface

Rapid technological developments in nanotechnology over the last decade have brought the field of therapeutics and diagnostics into a new area defined as nanomedicine. Nanotechnological applications are being used in biomedical areas such as therapeutics, diagnostics, monitoring, visualization, tissue engineering and even surgery. This cutting-edge technology uses the latest experimental approaches from life sciences, chemistry, physics, engineering and biology to solve problems in biomedicine, biotechnology and pharmaceutical industries.

This book provides an overview of the nanomedicine market, including descriptions of approved products and products in clinical trials. It also covers several aspects of drug delivery and tissue regeneration using different platforms, from biological bases like phages and peptides to nonbiological bases like lipids and polymers. In addition, it examines the use of microfluidic devices as advanced diagnostic techniques.

Islam Ahmed Hamed Khalil
Department of Pharmaceutics, College of Pharmacy and Drug Manufacturing,
Misr University of Science and Technology (MUST),
Giza, Egypt

Nanoscience Department, University for Science and Technology,
Zewail City of Science and Technology,
Giza, Egypt

Section 1

Introduction on Nanomedicines

Chapter 1

Introductory Chapter: Overview on Nanomedicine Market

Islam Ahmed Hamed Khalil, Islam A. Arida and Mohamed Ahmed

1. Introduction

Nanomedicine is an emerging field that has caught the interest of many medical scientists and chemists due to its unique characteristics that open the door wide for several unique applications that might lead to solving many problems that were found difficult to tackle in medicine. Nanomedicine has opened a new category of medicines called nanomedicines where the medicine is reduced to the nanoscale size, hoping to enhance its physicochemical properties. The chapter summarizes the nanomedicines that have been approved by the Food and Drug Administration (FDA) or European Medicines Agency (EMA) and the nanomedicines whose clinical trials based on previously published review articles by Anselmo and Mitragotri are ongoing [1, 2].

To gain insight to current trends in nanomedicine research and the most successful types of nanomedicines in the market, the approved nanomedicines are presented in **Figure 1**. The number of approved nanomedicine products is 29 till 2019 [1]. Liposomes represented 44.8% (13 products). Inorganic nanoparticles ranked second with 41.4% (12 products). Other nanoparticles (polymeric and protein) have only 4 products (13.8%). These findings are very interesting as liposomes are one of the oldest nanomedicines. This opens an argument about the challenges in nanomedicine translation as a new platform requires further investigations to prove its activity and safety. On the other hand, cancer nanotherapeutics is ranked first with 10 products in the market, followed by iron-replacement therapies with 8 products. Also, it is worth to mention that imaging agents (six marketed products) are ranked in third place, especially the inorganic nanoparticles (three products).

Moreover, nanomedicines, currently undergoing clinical trials, are presented in **Figure 2**. The number of products under clinical trials is 47 till 2019 [1], where liposomes represented 61.7% (29 products) and micelles ranked second with 19.15% (9 products). Other nanoparticles have only nine products (19.15%). Also, these findings are similar to approved nanomedicine, where liposomes are the most used nanomedicine. On the other hand, 39 products are dedicated for cancer treatment. It is worth to mention that 10 products out of the 39 products are loaded with gene therapy and not chemotherapeutic agents.

Generally, the total number of nanomedicines in the market or in clinical trial are 76 products, where liposome formulations were the most used delivery system with 55.26% (42 products), followed by inorganic nanoparticles with 21% (16 products) as presented in **Figure 3**. According to the World Health Organization in 2015, the first leading cause of death in around 50% of countries is cancer [3]. According to the International Agency for Research on Cancer report that published in 2018 on the global burden of cancer, there are 18.1 million cancer cases and 9.6 million cancer deaths in 2018 [3]. These reports inspired the

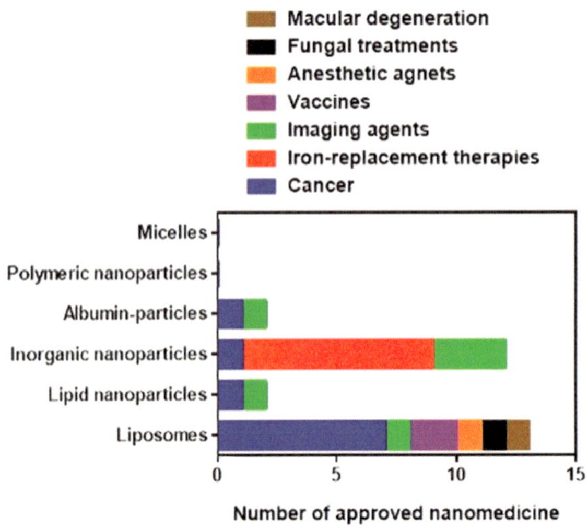

Figure 1.
Clinically approved nanomedicine for therapy and diagnostic.

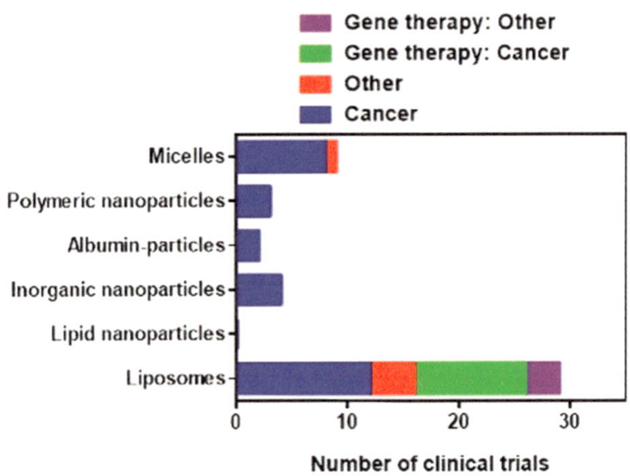

Figure 2.
Nanomedicine currently undergoing clinical trials for therapy and diagnostic.

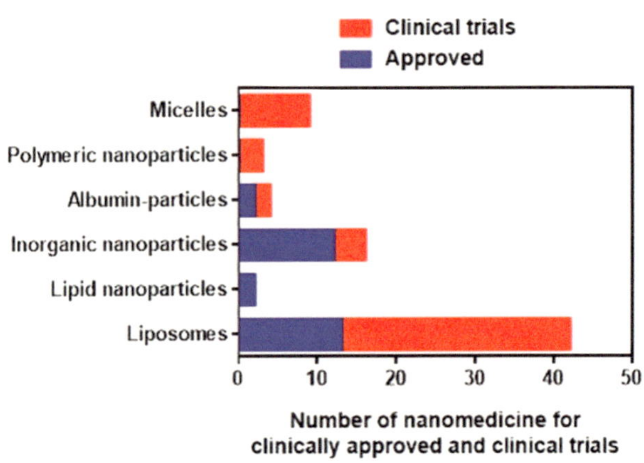

Figure 3.
Total number of nanomedicines approved and under clinical trials (therapeutic and diagnostic).

pharmaceutical industry to invest in this market. As mentioned previously, there are only 10 nanomedicine products out of 29 products available in the market to treat cancer, while there are 39 products out of 47 products for cancer treatment. This number of clinical trials for cancer is mainly derived by the 15 years of support by the US National Cancer Institute through the Centers of Cancer Nanotechnology Excellence (CCNEs) [4].

2. Types of nanomedicines

Nanomedicines are mainly classified into two classes, either inorganic nanoparticles such as gold, silica, and iron oxide or organic nanoparticles such as polymeric, liposomes, and micelles (**Figure 4**). These nanoparticles are mostly used for therapeutic and diagnostic nanoparticles. Inorganic nanoparticles have been used for a variety of applications including lymph node imaging, hyperthermia, and anemia treatment. Some of them have successfully gone through preclinical studies and clinic trials. Along with inorganic nanoparticles, organic-based nanoparticles have successfully reached the clinical phase and currently reached the market for different applications like vaccination, microbial infection, and cancer.

2.1 Liposome-based nanomedicines

Liposome-based nanomedicine is a type of drug formulation where a drug is encapsulated inside the phospholipid bilayer structure to enhance its bioavailability and therapeutic activity. Liposome formulations are one of the oldest nanomedicines with a well-established technique. Many research efforts were focused on using liposomes to encapsulate several cargos like small molecules such as doxorubicin, nucleic acid such as RNAs, and biological molecules such as vaccines for hepatitis A virus. Furthermore, administration of the liposomes without an encapsulated drug is also a possibility if the liposome subunits have a certain therapeutic effect such as sphingomyelin and cholesterol. PEGylation is an option to consider while using liposomes due to its importance in adding stealth to the delivery system. Most of the approved liposome-based nanomedicines are used for the treatment of cancer diseases. They take a large place in research as 10 out of the 29 approved nanomedicines are liposome-based.

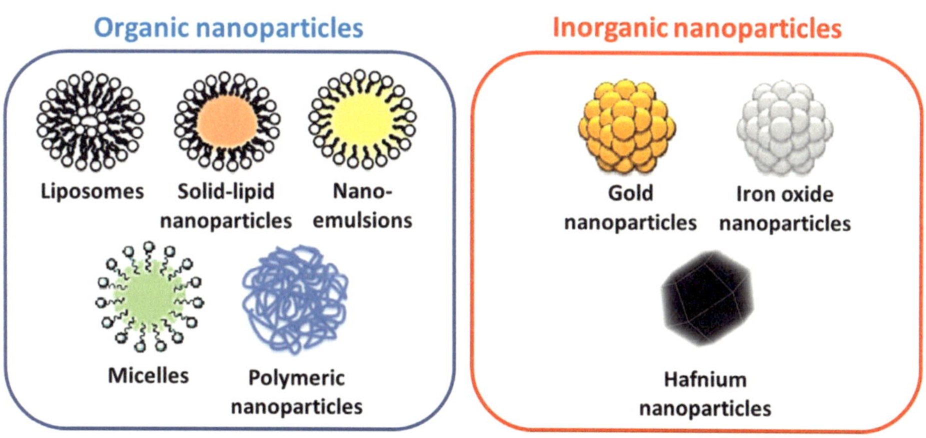

Figure 4.
Clinically approved and investigated nanomedicines including organic nanoparticles and inorganic nanoparticles.

2.2 Lipid-based nanomedicines

Lipid nanosystems including nanoemulsions and solid lipid-based nanoparticles are another form of nanomedicine, which are usually used to encapsulate hydrophobic cargos to improve permeation and control release profile. Usually, a surfactant is used to ensure a uniform dispersion. Lipid nanomedicine can also encapsulate some gene therapeutics such as siRNA or contrast agents used for imaging such as F-butane. Generally, lipid nanomedicine can improve the pharmacological effect by enhancing drug accumulation in targeted tissues beside its biocompatibility. However, there are several drawbacks like rapid clearance due to reticuloendothelial system (RES) uptake and some limitations for administration routes and challenges regarding system stability [5, 6]. Unlike liposomal-based nanomedicines, lipid-based nanomedicines are not limited for cancer diseases only. Some of the diseases that are treated by lipid-based nanomedicines are amyloidosis, hepatitis B, and hepatic fibrosis. Furthermore, several types of nanoemulsion were loaded with drugs like simvastatin, cinnarizine, coenzyme Q10, and cyclosporine, which used as antihyperlipidemia, antihistaminic, antioxidant, and immunosuppressants, respectively.

2.3 Albumin-based nanomedicines

Albumin-based nanomedicines are another form of nanosystems, where albumin, especially human serum albumin (protein), is used as a carrier. Albumin nanosystems can be loaded with different cargos via a simple self-assembly procedure of albumin in aqueous solution with simple crosslinking step. The main advantage of albumin is biocompatibility. Despite that, only 2 out of the 29 listed approved nanomedicines and 2 out of the 65 nanomedicines under clinical trials are albumin-based. It is currently used in imaging and delivering drugs that treat cancer diseases.

2.4 Micelle-based nanomedicines

Micelles are self-assembled nanosystem by amphiphilic molecules that have a hydrophilic part and a hydrophobic one. They have several advantages like high permeability and solubility, which improve drug bioavailability. However, they still have some drawbacks like insufficient control to drug release and cytotoxicity due to amphiphilic molecule use, which interact with cell membrane [5, 7]. Although several reports used block copolymeric micelles to reduce clearance and increase bioavailability of chemotherapeutic agents and other types of drugs, there are no approved micelle-based nanomedicines. However, there are currently nine micelle-based nanomedicines undergoing clinical trials. Majority of them are used for cancer treatment.

2.5 Polymeric-based nanomedicines

Polymeric nanoparticles are one of the most commonly used nanosystems for drug delivery. Several polymers have been used like ethyl cellulose, poly(lactic-co-glycolic acid), polylactic acid, cyclodextrin, alginate, and chitosan. Depending on the nature of the polymer, either hydrophilic or hydrophobic, there are several techniques that have been used to prepare polymeric nanoparticles. Several advantages like relative stability and prolonged duration of action make polymeric nanoparticles a promising platform for the market. However, there are no marketed products based on polymeric nanoparticles. Only three products are currently on clinical trials for cancer.

2.6 Inorganic-based nanomedicines

Inorganic-based nanomedicines have several subtypes. Due to degradability and biocompatibility issues, few types have been used for therapeutic purpose, while other types for diagnostic purpose like imaging agents. One of these subtypes are metal oxide nanoparticles such as hafnium oxide nanoparticles which enhance tumor cell death via electron production through their stimulation with external radiation. Another subtype is in the form of colloids such as iron dextran colloids, iron gluconate colloid, and other similar derivatives that are usually used for the treatment of iron-deficiency anemia. The last subtype mentioned is iron–/silica–/gold-based nanomedicines, either as nanoparticles with drugs arranged on the surface for the treatment of cancer or as nanoshells/nanoparticles used for thermal ablation of tumors. There are 12 products in the market that belong to this type. Eight products used iron-replacement therapies. On the other hand, four products are currently on clinical trials for treating cancer.

3. Nanomedicines pharmacokinetic and regulations

The pharmacokinetic parameters of nanomedicines are similar to free drugs with addition phase after drug administration, which is the liberation phase beside the standard absorption, distribution, metabolism, and excretion (ADME). This new phase is controlled by particle nature, size, shape, and surface properties. It is worth to mention that particle size is very important for absorption and elimination. Particles with particle size <5 nm is easily excreted from the kidney, while larger particle size could be eliminated by the liver or engulfed by mononuclear-phagocyte system. Moreover, particle size and shape can affect particle accumulation in targeted tissues like ellipsoidal shape that has better distribution and retention in tumor tissue than spherical one. Surface modification of nanoparticles can affect particle uptake and elimination. Many nanoparticles are coated for active and passive targeting. Passive targeting is a non-specific retention in target tissue like solid cancer tissue by enhanced permeability and retention mechanism. Active targeting is the selective uptake of nanomedicine by specific cells. Target moieties could be protein, antibody, or small molecule selective to specific tissues or cells. This mechanism is mainly controlled by homing to overexpressed cell surface receptors.

The Food and Drug Administration classified nanoscale materials to nanomaterials as "materials used in the manufacture of nanomedicine" or nanomedicine as "final products," The FDA approved 51 nanomedicines by the year 2016, 40% of which were in clinical trials between 2014 and 2016. According to the FDA evaluation of nanomedicines, it includes the physicochemical properties, followed by pharmacokinetics evaluation of nanomedicines. The pharmacokinetics evaluation includes (1) rate and amount of absorption, (2) retention in circulation, (3) half-life and complete elimination, (4) bioavailability differences, (5) distribution or accumulation to the body or specific tissue for active targeting, (6) decomposition or metabolism, (7) elimination, and (8) toxicity assessment of nanomedicines. On the other hand, the European Medicines Agency defined nanomedicines as "drugs composed of nanomaterials 1–100 nm in size, and these are classified into liposomes, nanoparticles, magnetic nanoparticles, gold NPs, quantum dots, dendrimers, polymeric micelles, viral and non-viral vectors, carbon nanotubes, and fullerenes." EMA has approved eight commercially available nanomedicines as first-generation nanomedicines. Currently, there are 48 nanomedicines in clinical trials (Phases 1–3) in the EU. EMA evaluates the pharmacokinetics and pharmacodynamics of nanomedicines through investigation of their chemical composition and physicochemical properties [8].

4. Approved application and indication of nanomedicine

4.1 Cancer nanoparticle medicines

Most pharmaceutical industries are focusing on developing new products for cancer as it is the first cause of death in 50% of the countries. Nanomedicine products have a good share in this market with many approved products to treat several types of cancer at various stages. Abraxane® is a famous albumin-particle bound paclitaxel nanomedicine loaded for advanced non-small cell lung cancer, metastatic breast cancer, and metastatic pancreatic cancer. Doxil®, the first approved nanomedicine by the FDA in 1995, is a PEGylated liposome loaded with doxorubicin for ovarian cancer, HIV-associated Kaposi's sarcoma, and multiple myeloma. Marqibo® is a liposomal vincristine for Philadelphia chromosome-negative acute lymphoblastic leukemia. Hensify® is the recently approved nanomedicine for cancer in 2019 by the FDA. It is the hafnium oxide nanoparticles stimulated with external radiation to enhance tumor cell death via electron production for locally advanced squamous cell carcinoma. Most of the approved nanomedicines are non-PEGylated except Doxil and Onivyde, which is interesting as most reports have proven the importance of nanomedicine coating with PEG. Furthermore, all nanomedicine products do not have active target moiety. So, all of these products follow passive targeting approach without even stealth characteristics.

4.2 Iron-replacement nanoparticle therapies

Iron-replacement therapy to treat anemia is surprisingly another area for nanomedicine due to the significance of nanoscale iron-oxide colloid system in improving iron absorption to the body. The main advantage of iron-oxide nanomedicine is replacing the injection of free iron with its associated toxicity. Most of these nanosystems are coated with either polysaccharide or polymer to reduce iron toxicity. CosmoFer® is the first approved iron dextran colloid by the FDA in 1996. Injectafer® is the most recent one in 2013 by the FDA, which is iron carboxymaltose colloid.

4.3 Nanoparticle/microparticle imaging agents

Another area for nanomedicine, especially the inorganic ones, is diagnostics, mainly imaging agents. Iron-oxide nanomedicines are also approved as contrasting agents for magnetic resonance imaging, which is used to generate contrasted images for different types of cancers. The magnetic property and small particle size allow the distribution of iron-oxide nanomedicine in tumor tissue, which provide a precise imaging of cancer borders. Additionally, perflutren is also used as ultrasound contrast agent in either lipid- or albumin-based nanomedicines. Phospholipid-stabilized microbubble is another form of nanomedicine as ultrasound contrast agent, which is approved in 2001 by the EMA. Its main mechanism is encapsulating air bubbles, which act as reflectors for ultrasound.

4.4 Nanoparticles for vaccines, anesthetics, fungal treatments, and macular degeneration

Several clinical applications have been studied using nanomedicine. Diprivan® is the first FDA-approved nanomedicine in 1989 for anesthesia. Another field for nanomachine is vaccination with two products, which are Epaxal® for hepatitis A and Inflexal V® for influenza. Both vaccines are liposome-based nanomedicine due to the similarity of liposome structure to cell structure. Another famous liposome product is AmBisome®, which is a liposome loaded with amphotericin B for treating systemic fungal infections with reduced toxicity. Abelcet® is another approved lipid-based nanomedicine loaded with amphotericin B. Finally, Visudyne® is a liposomal verteporfin for treatment of subfoveal choroidal neovascularization from age-related macular degeneration, pathologic, or ocular histoplasmosis.

5. Conclusion

Nanomedicines are currently in the middle of the road with great potentials but require many development considerations regarding assessment of physicochemical properties, pharmacokinetic properties, and pharmacodynamic applications. Based on the recent trends with 47 products in clinical trial phases, it is expected that within the next few years, more products will be available for several applications, especially cancer.

Conflict of interest

The authors declare no conflict of interest.

Author details

Islam Ahmed Hamed Khalil[1,2]*, Islam A. Arida[2] and Mohamed Ahmed[2]

1 Department of Pharmaceutics, College of Pharmacy and Drug Manufacturing, Misr University of Science and Technology (MUST), Giza, Egypt

2 Nanoscience Department, University for Science and Technology, Zewail City of Science and Technology, Giza, Egypt

*Address all correspondence to: islam.khalil@must.edu.eg

References

[1] Anselmo AC, Mitragotri S. Nanoparticles in the clinic: An update. Bioengineering & Translational Medicine [Internet]. 2019;**4**(3):1-16. DOI: 10.1002/btm2.10143

[2] Anselmo AC, Mitragotri S. Nanoparticles in the clinic. Bioengineering & Translational Medicine. 2016;**1**(1):10-29

[3] Bray F, Ferlay J, Soerjomataram I, Siegel RL, Torre LA, Jemal A. Global cancer statistics 2018: GLOBOCAN estimates of incidence and mortality worldwide for 36 cancers in 185 countries. CA: A Cancer Journal for Clinicians [Internet]. 2018;**68**(6):394-424. DOI: 10.3322/caac.21492

[4] Park K. The beginning of the end of the nanomedicine hype. Journal of Controlled Release [Internet]. 2019;**305**:221-222. DOI: 10.1016/j.jconrel.2019.05.044

[5] Kawabata Y, Wada K, Nakatani M, Yamada S, Onoue S. Formulation design for poorly water-soluble drugs based on biopharmaceutics classification system: Basic approaches and practical applications. International Journal of Pharmaceutics [Internet]. 2011;**420**(1):1-10. DOI: 10.1016/j.ijpharm.2011.08.032

[6] Mora-Huertas CE, Fessi H, Elaissari A. Polymer-based nanocapsules for drug delivery. International Journal of Pharmaceutics [Internet]. 2010;**385**(1-2):113-142. Available from: https://linkinghub.elsevier.com/retrieve/pii/S0378517309007273

[7] Devalapally H, Chakilam A, Amiji MM. Role of nanotechnology in pharmaceutical product development. Journal of Pharmaceutical Sciences [Internet]. 2007;**96**(10):2547-2565. DOI: 10.1002/jps.20875

[8] Choi YH, Han H-K. Nanomedicines: Current status and future perspectives in aspect of drug delivery and pharmacokinetics. Journal of Pharmaceutical Investigation [Internet]. 2018;**48**(1):43-60. DOI: 10.1007/s40005-017-0370-4

Section 2

Biological Based Nanomedicines

Chapter 2

Phage Capsids as Gated, Long-Persistence, Uniform Drug Delivery Vehicles

Philip Serwer, Elena T. Wright and Cara B. Gonzales

Abstract

Over the last 25 years, cancer therapies have improved survivorship. Yet, metastatic cancers remain deadly. Therapies are limited by inadequate targeting. Our goal is to develop a new drug delivery vehicle (DDV)-based strategy that improves targeting of drug delivery to solid tumors. We begin with a capsid nanoparticle derived from bacteriophage (phage) T3, a phage that naturally has high persistence in murine blood. This capsid has gating capacity. For rapidly detecting loading in this capsid, here, we describe procedures of native agarose gel electrophoresis, coupled with fluorescence-based detection of loaded molecules. We observe the loading of two fluorescent compounds: the dye, GelStar, and the anticancer drug, bleomycin. The optimal emission filters were found to be orange and green, respectively. The results constitute a first milestone in developing a drug-loaded DDV that does not leak when in blood, but unloads its cargo when in a tumor.

Keywords: agarose gel electrophoresis, bacteriophage T3, bleomycin, buoyant density centrifugation, capsid impermeability, GelStar

1. Introduction

Current therapies for cancerous tumors suffer from both toxic secondary effects and the development by the tumor of drug resistance. These effects usually block therapy for metastatic cancers, the cause of 90% of cancer deaths [1–5]. For solving these problems, our first thesis is that the best strategy is to increase tumor specificity of anticancer drug delivery in several, *independent* stages. If, for example, three stages are used and each stage is 80% efficient (20% nonefficient) in increasing specificity, then overall efficiency is 99% [$100 \times (1.0 - 0.2^3)$]. In this case, (1) drug dosages to tumors can be raised 100× without changing toxicity and, therefore, (2) tumor cell evolution of drug resistance is minimized.

The primary alternative is to continue testing chemotherapies [6–8], immunotherapies [9–11] and radiotherapies [12–14] that have tumor-specificity determined at one independent stage. This one stage is often cellular DNA replication, which is more rapid and, therefore, more drug- and radiation-sensitive, in cancerous cells than it is in healthy cells. One-stage strategies are >100 years old for immunotherapy and radiotherapy. Chemotherapeutic agents typically used are over 50 years old [8]. Even major effort has not produced systematic therapies for metastatic cancer.

IntechOpen

Apparently, new, possibly more biology-based, strategy is needed to counter risk that the above one-stage-based strategies, in general, are not realistic for reaching objectives (reviewed in [1, 15–17]).

In theory, one implementation of multi-stage strategy starts with drug delivery in a drug delivery vehicle (DDV) that is gated. The gate is opened to load drug, closed in circulation to deliver drug and opened again in tumors to administer drug. Stages of DDV-derived toxicity reduction are the following tumor-specific events: (1) DDV delivery, (2) DDV opening and, (3) drug activation, for masked drugs. We address stages (1) and (2) here.

In this implementation, tumor-specific delivery is achieved via the EPR effect. The EPR effect is the spontaneous accumulation of nanoparticles in tumors, observed for an uncharacterized phage in 1940 [18] (reviewed in [19]). The causes of the EPR effect are (1) porosity of tumor blood vessels and relative tightness of healthy blood vessels, so that nanoparticles enter tumors, but usually not healthy tissue and (2) poor tumor lymphatic drainage, so that nanoparticles remain [20–25]. The EPR effect is the basis for the use of several FDA-approved, drug-loaded, liposomal DDVs [23, 24, 26].

However, circulating, drug-loaded, liposomal DDVs undergo drug leakage that causes significant toxicity [22, 23, 25, 27]. Also, liposomes are removed from circulation by the macrophage-phagocyte innate immune system. That is to say, liposomes are not very persistent. Chemical solutions to the leakage problem do not exist to our knowledge. Chemical solutions to the persistence problem (e.g., polyethylene glycol derivatization [28, 29]) introduce quality control problems and are not adaptable to future improvements, for example, achieving of tumor-specific unloading. The second thesis is that the optimal solution is linked to finding an appropriate, biologically produced, microbial DDV. Unlike some biology-based anticancer strategies [1, 15, 16, 19], use of a DDV is implementable with nononcolytic viruses and, therefore, avoids the dangers [19] of using oncolytic viruses.

In practice, we have discovered a phage T3 capsid that appears to have DDV-favorable characteristics needed for implementation of our strategy. First, phage T3 (and presumably its capsid) has recently been found to have exceptionally high persistence in mouse blood (3–4 h), unlike the T3 relative, T7 [30]. Second, one empty, but otherwise phage-like T3 capsid, is impermeable (for over 20 years) to the compound, Nycodenz (821 Da molecular weight) (reviewed in [31]). But, when the temperature is raised to 45°C, Nycodenz enters this capsid [32], presumably through a gate that opened. The concept is that, if we adapt this capsid to use as gated DDV, some of the needed engineering has already been done by natural selection.

Phages T3 and T7 are illustrated at the left in **Figure 1**. The gated capsid is illustrated at the right in **Figure 1**. This capsid is generated during DNA packaging that had been initiated by a DNA-free procapsid called capsid I (not shown). During packaging, capsid I expands and becomes the more angular and stable capsid (capsid II) illustrated in **Figure 1**. In nature, the capsid-gate is a ring of 12 gp8 molecules (**Figure 1**) that acts as entry portal for DNA during DNA packaging [31–33]. Most T3 and T7 capsid II particles are purified after having detached from DNA during infected cell lysis. The amount is 5–10 mg capsid II per liter of culture. The last purification step is buoyant density centrifugation in a Nycodenz or Metrizamide density gradient. Nycodenz low density (NLD) capsid II has the gp8 "gate" and is impermeable to Nycodenz and Metrizamide. The low density (1.08 g/ml) is caused by high internal hydration, which is caused by Nycodenz impermeability.

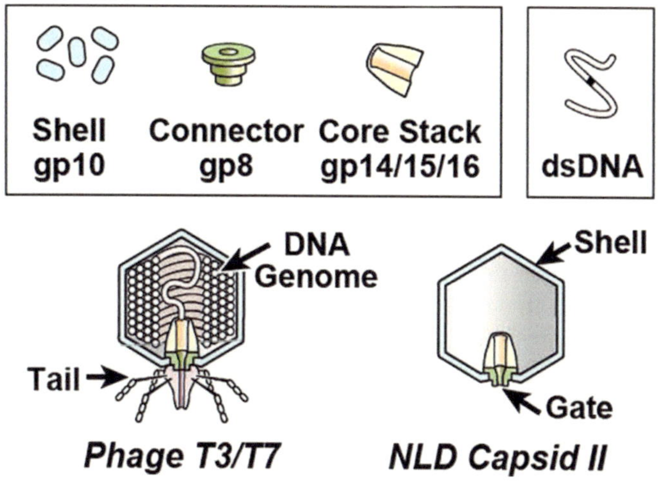

Figure 1.
Phages T3 and T7 (left); and T3 and T7 NLD capsid II (right). The graphic legend at the top indicates the various capsid components. A protein is labeled by gp, followed by the number of the encoding gene. Analogous T3 and T7 genes have the same numbers. If NLD capsid II is used as DDV, the perimeter of the DDV is defined by the gp10 shell; the gate is the gp8 portal. All structures are the same for phage T3 as they are for phage T7 (reviewed in [31, 32]).

Nycodenz high density (NHD) capsid II is Nycodenz-permeable (1.28 g/ml) and is completely separated from NLD capsid II during buoyant density centrifugation in a Nycodenz or Metrizamide density gradient [31, 33–35]. We use NHD capsid II as a control during loading experiments.

In the case of T3 NLD capsid II, gated entry of Nycodenz has been observed via raising of temperature. The likely entry channel was the axial hole of the gp8 ring. This conclusion was drawn, in the case of T7 NLD capsid II, from (1) entry kinetics of the fluorescent dye, bis-ANS [1,1′-bi(4-anilino)naphthalene-5,5′-di-sulfonic acid; 673 Da], and (2) covalent cross-linking of bis-ANS to channel proteins [33]. Asymmetric reconstruction-cryo-EM [36] revealed that obstruction of the T7 gp8 channel (and presumably the T3 channel) varies.

In support of working to implement the above gating-based strategy, the following quote from 2005 presents an expert opinion of what is needed for the next generation of anticancer DDVs [27]. "An ideal liposomal anticancer drug would exhibit little or no drug release while in the plasma compartment, thus ensuring limited exposure of the drug to healthy tissue. This feature would also maximize drug delivery to disease sites, as mediated by the movement of the drug-loaded liposomes from the plasma compartment to the extravascular space at disease sites, such as a region of tumor growth. Following localization, however, the drug-loaded liposome must transform itself from a stable carrier to an unstable carrier. This would ensure that the drug, which has localized in the diseased site, is bioavailable." To our knowledge, no details for such "controlled release" via a liposomal DDV have been published. The system described here is designed to accomplish what is described in the above quote. However, implementation uses gating of a DDV, not programmed instability of a DDV.

To proceed further, we need a procedure for rapidly determining whether a drug is loaded in a capsid. In the current study, we have developed native agarose gel electrophoresis (AGE) for this purpose. The detection is performed via capsid band fluorescence produced by the compound loaded.

2. Results

2.1 Detection of test compounds: GelStar

We tested the loading of two fluorescent compounds. The first was GelStar, a fluorescent nucleic acid stain typically used after AGE. In contrast, we incubated GelStar with our capsid and then performed AGE without further use of GelStar. The second compound was bleomycin, an anticancer drug [37, 38] that is also fluorescent [38]. Neither the manufacturer nor the vendor provided either the structure of GelStar or the concentration of commercial GelStar solutions. GelStar is sold in solution only.

The dominant fluorescence emission of nucleic acid-bound GelStar is in the green range. Apparently not previously documented is that the dominant fluorescence emission of free GelStar is in the orange range, at least when the GelStar is in an agarose gel. Ultraviolet light stimulated GelStar fluorescence emission vs. GelStar dilution is shown in **Figure 2**. Free GelStar, at several dilutions, had been pipetted in 5 μl amounts onto an agarose gel before ultraviolet light illumination and photography through an orange filter (spots labeled G in **Figure 2**). The effective volume in μl (dilution, multiplied by 5) of the stock GelStar solution is also indicated. In **Figure 2**, the color of GelStar spots is orange for all dilutions, as it also is found to be (not shown) with yellow and green emission filters. The orange color is real and is not produced by the emission filter because green Alexa 488 dye fluorescence retains its green color (spots labeled A in **Figure 2**). The number next to the Alexa 488 spots is the total amount (μg) of Alexa 488, also applied in 5 μl amounts.

DNA-bound GelStar had the expected green emission at all dilutions, when viewed through the same orange emission filter used for **Figure 2** (right side of the right panel of **Figure 3**). Green emission was also dominant when yellow and green emission filters were used for DNA-bound GelStar (not shown).

2.2 Detection of test compounds: bleomycin

Without fluorescent compound, the background of an agarose gel was blue when emission was photographed without an emission filter (not shown). Thus, not surprising was that optimal detection of bleomycin was not obtained with a blue filter, even though the blue range was where peak emission was previously found for bleomycin [38]. Among the blue, green, yellow and orange filters, optimal detection was obtained with the green filter.

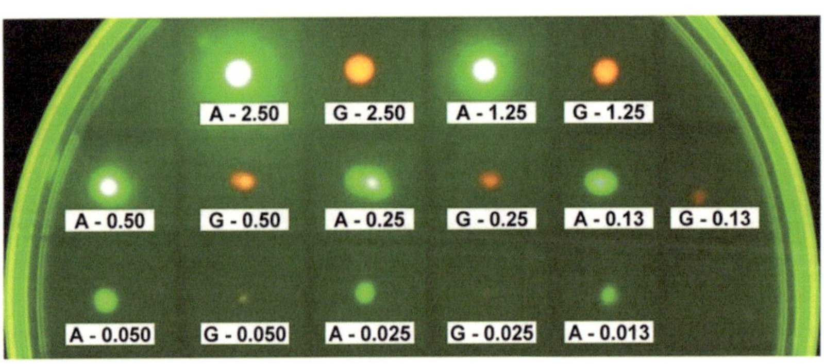

Figure 2.
Fluorescence of free GelStar. GelStar and Alexa Fluor 488 were diluted and, then, pipetted onto the surface of an agarose gel. The fluorescence was photographed through the orange filter. The GelStar samples are indicated by G, followed by the effective volume (μl) of the original, undiluted GelStar solution. The Alexa Fluor 488 samples are indicated by A, followed by the amount (μg) of Alexa Fluor 488.

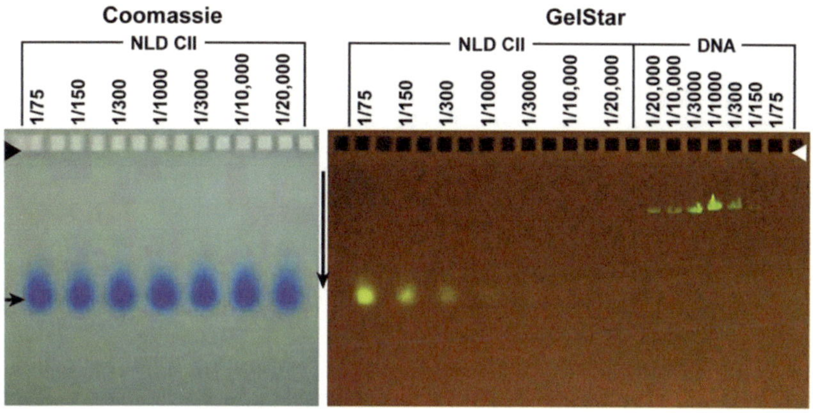

Figure 3.
Association of GelStar with T3 NLD capsid II. A commercial GelStar solution was diluted to the extent indicated above a lane and incubated with T3 NLD capsid II. Association of GelStar with the capsid was determined by AGE, followed by, first, photography of fluorescence (right panel, lanes labeled NLD CII) and, then, staining of protein with Coomassie blue (left panel). Also analyzed was purified T3 DNA (right panel, lanes labeled DNA), which does not stain with Coomassie blue. The arrow indicates the direction of electrophoresis; the arrowheads indicate the origins.

With the green emission filter, the minimal detected bleomycin amount was 0.2–0.4 ng when a bleomycin dilution series like the GelStar dilution series in **Figure 2** was photographed (not shown). Contrast enhancement of images was used at these lower amounts.

2.3 Loading of GelStar in NLD capsid II

We succeeded in loading GelStar into NLD capsid II. To achieve loading, 10 μg of NLD capsid II was incubated with GelStar at 45°C. Loading was then assayed by AGE at 10°C. Then, the gel was illuminated with ultraviolet light. The result was a fluorescent band of intensity that monotonically increased with decreasing GelStar dilution (left section of the right panel of **Figure 3**). The capsid amount was invariant, as judged by Coomassie staining of the same gel (left panel of **Figure 3**). At GelStar dilutions lower than those in **Figure 3**, down to 1/10, the band intensity reached a plateau (not shown). The dominant fluorescence, at all dilutions, was green, implying that the GelStar was bound to something capsid associated. GelStar did not detectably associate with NHD capsid II (not shown).

The following data indicated that the GelStar-binding capsid site was not on a DNA molecule associated with the capsid. As the dilution of GelStar decreased, the DNA-bound GelStar fluorescence underwent, first, an increase and then a decrease (**Figure 3**, right segment of right panel). However, the decrease was not observed for the binding to NLD capsid II. Second, although a minor NLD capsid II fraction has DNA [31], the DNA-containing NLD capsid II had been excluded during purification by selecting the low-density side of the NLD capsid II band after buoyant density centrifugation in a Nycodenz density gradient. Thus, the GelStar was apparently either self-bound or bound to capsid protein.

2.4 Loading of bleomycin in NLD capsid II

Association of bleomycin with T3 NLD capsid II was also achieved. However, the fluorescence signal was relatively weak (**Figure 4**). The bleomycin fluorescence signal of a NLD capsid II band did not change when the concentration of bleomycin was changed from 2 to 16 mg/ml. A bleomycin-associated NLD capsid II band is

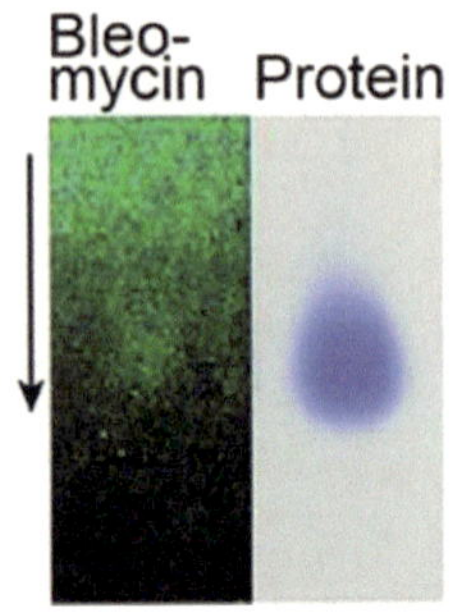

Figure 4.
Association of bleomycin with T3 NLD capsid II. The experiment of ***Figure 3*** *was repeated with bleomycin (8 mg/ml), instead of GelStar. The capsid region of the post-AGE gel is shown. The right (protein) panel has a single band of capsid stained with Coomassie blue. This band marks the position of the capsid-associated bleomycin fluorescence in the left panel. The arrow indicates direction of electrophoresis.*

shown in **Figure 4**. Most of the free bleomycin migrated toward the cathode (not shown), i.e., in a direction opposite to the direction of capsid migration.

The strength of the signal in **Figure 4** was weakened by the blue background and use of a green filter. In addition, a contaminant in the bleomycin preparation migrated close to the capsids, and is seen above the capsid band at the top of the left panel of **Figure 4**.

Calibration data for bleomycin, like the data for GelStar in **Figure 2**, were obtained. These data revealed that the amount of bleomycin loaded was 150–300 molecules per capsid.

3. Discussion

In the Introduction, we outlined a strategy that is expected to work, if we can achieve the following objectives: (1) high (~4 h) persistence of NLD capsid II in blood so that the EPR effect has time to work, (2) adequate loading and sealing of NLD capsid II and (3) tumor-specific, controlled release (de-sealing or unloading). Objective #1 is likely already achieved, given the high persistence of T3 phage. That is to say, if one considers this strategy to be engineering based, some of the engineering might already be been done by natural selection.

Concerning adequate loading, the volume of the internal cavity of NLD capsid II = 6.95×10^{-17} ml. For volume occupancy (F_V) of 0.5 (equal to the F_V of DNA packaged in mature phage [34]), the number of bleomycin molecules (1416 Da; density estimated at 1.6 g/ml as sulfate) per NLD capsid II particle is 2.4×10^4. The recommended dose of DDV-free bleomycin depends on the tumor, but is typically [39, 40] 10–20 units/m^2, corresponding roughly to 10–20 mg/m^2; 15 mg/m^2 is 1.76×10^{18} bleomycin molecules/m^2.

To calculate the number (N_D) of NLD capsid II particles needed for this dose at $F_V = 0.5$, we initially assume a 25 g mouse, which on average, has 78.6 cm^2 surface area [41]. Then, N_D is 3.6×10^{11}. A 6-liter culture yields 150–300 mouse doses of this size (cost ~ $1500), assuming (1) laboratory-scale production technique, (2) no development of procedures to increase the amount produced per bacterial cell, and (3) no drug-dose reduction caused by improved targeting. That is to say, if we can half-fill the volume of NLD capsid II, we have a viable beginning. However, thus far, we have filled no more than 2% of $F_V = 0.5$, NLD capsid II volume. So, increasing the loading is a major objective for the future.

An apparent obstacle to achieving this goal is the nonincrease in loading as bleomycin concentration increases above 2 mg/ml. At least two possible explanations exist. (1) After passing through an open gate, the bleomycin eventually causes the gate to close. We were hoping to close the gate by lowering temperature. (2) After diffusing through an open gate, the bleomycin is prevented from diffusing in reverse by binding to internal proteins; the internal proteins become saturated as the concentration of bleomycin increases. In either case, increasing the loading is a problem of engineering.

An advantage of using a phage DDV is that the human-design engineering potential is relatively high. First of all, the capsids in a T7 NLD capsid II preparation are structurally uniform enough so that symmetric cryo-EM reconstruction is obtained at 3.5 Å [34] and asymmetric reconstruction, at ~8 Å [36]. Assuming T3 capsids to be comparably homogeneous, use of chemistry to improve gating should produce relatively uniform results.

Second, phages, in general, and phage T3 in particular, can be genetically manipulated, which is not possible with liposomes. Information for determining which nucleotides to change can be obtained from high-resolution cryo-EM structure. Structure of this type is not obtainable with liposomes.

Finally, we note that, as far as we know, the only phages tested for production of an NLD capsid II-like capsid are the related coliphages, T7, T3 and ϕII. All three of these phages produce a NLD capsid II-like capsid [42]. Other phages are potential sources of gated capsids, perhaps with properties even more DDV-favorable.

4. Materials and methods

4.1 T3 bacteriophage, capsids and DNA (nanoparticles)

We obtained bacteriophage T3 and T3 capsid II from 30°C-lysates of host, *Escherichia coli* BB/1, that had been infected by phage T3 in aerated liquid culture [43]. The growth medium was 2× LB medium: 2.0% Bacto tryptone, 1.0% Bacto yeast in 0.1 M NaCl. We initially purified both phage and capsids by centrifugation through a cesium chloride step gradient, followed by buoyant density centrifugation in a cesium chloride density gradient [43]. The latter fractionation separates capsid I from capsid II.

To separate NLD capsid II from NHD capsid II, we performed buoyant density centrifugation of capsid II in a Nycodenz density gradient, as previously described [32]. The purified NLD and NHD capsid II were dialyzed against 0.1 M NaCl, 0.01 M Tris-Cl, pH 7.4, 0.001 M $MgCl_2$. NLD capsid II, which formed a band near the top of the Nycodenz density gradient, had no detected contamination with NHD capsid II and vice versa, as previously seen by analytical ultracentrifugation [31]. Phage, NLD capsid II and NHD capsid II were dialyzed against the following buffer before use in the experiments described below: 0.2 M NaCl, 0.01 M Tris-Cl, pH 7.4, 0.001 $MgCl_2$.

T3 DNA was obtained from purified T3 phage by phenol extraction. The DNA was dialyzed against and stored in 0.1 M NaCl, 0.01 M Tris-Cl, pH 7.4, 0.001 M EDTA. DNA concentration was obtained from optical density at 260 nm.

4.2 Fluorescent compounds: test of fluorescence emission

GelStar was obtained from Lonza (Basel, Switzerland) in solution. The company recommends dilution by a factor of 1:10,000 for use as a nucleic acid stain after gel electrophoresis. Alexa Fluor 488 succinimidyl ester was obtained from Molecular Probes (Eugene, OR, USA) as a powder.

Bleomycin was obtained from Cayman Chemical Company (Ann Arbor, MI, USA) as a powder. The bleomycin was dissolved in the aqueous buffer indicated and diluted to the concentrations indicated before incubation with capsids and DNA.

Tests of fluorescence emission vs. fluorescent molecule concentration were made by pipetting 5 μl of diluted fluorescent molecule onto the surface of a 0.7% agarose gel (LE agarose, Lonza) that had been cast in a plastic Petri dish in the electrophoresis buffer of Section 4.3. The gel was then photographed by use of the procedures described in Section 4.3.

4.3 Loading experiments: AGE

To test for fluorescent compound/nanoparticle association, fluorescent compounds were mixed with one of the following T3 nanoparticles: NLD capsid II, NHD capsid II, phage, DNA. First, a 12.5 μl amount of fluorescent compound in 0.1 M NaCl, 0.01 M sodium citrate, pH 4.0, 0.001 M $MgCl_2$ (citrate buffer) was added to 4.5 μl of additional citrate buffer. Then, 8.0 μl of a nanoparticle sample was added and mixed (final pH, 4.1). This mixture was incubated at 45.0°C for 2.0 h.

To perform AGE, we added to this mixture 2.5 μl of the following solution: 60% sucrose (to increase the density for layering in sample wells) in the electrophoresis buffer below. This final mixture was layered in a well of a horizontal, submerged, 0.7% agarose gel (LE agarose, Lonza), cast in and submerged under the following electrophoresis buffer: 0.05 M Tris-acetate, pH 8.4, 0.001 M $MgCl_2$. The temperature of the gel and buffer had been pre-adjusted to 10°C in an effort to seal NLD capsid II and, therefore, prevent leakage of fluorescent compounds.

AGE was performed at 1.0 V/cm for 18.0 h with the gel and buffer maintained at 10°C by circulation through a controlled-temperature water bath. After AGE, the gel was soaked in 25% methanol in electrophoresis buffer for 1.5 h at room temperature, to cause leakage of fluorescent compounds from NLD capsid II and, therefore, to prevent auto-quenching.

Finally, the gel was photographed during illumination with a Model TM-36 ultraviolet transilluminator (Ultra Violet Products, Inc.). The camera used was a Canon Power Shot G2, 4.0 Megapixels. The following Tiffin emission filters were used as described in Section 2: Blue, 80A #290513; Green, 11Green 1—#287305; Yellow, Yellow 12—#282224; Orange, Orange 16—#289750. To detect capsid protein, the gel was subsequently stained with Coomassie blue and photographed during illumination with visible light [43].

5. Conclusions

Obtaining an increase in the current tumor-specificity of anticancer drugs should be possible via use of a DDV that implements multiple, independent stages of specificity increase. T3 NLD capsid II is an example of a bio-nanoparticle that has undergone some of the needed DDV-bioengineering via mutation/selection in the environment. Other examples, not yet found, are assumed to exist and potentially have even more favorable characteristics.

Acknowledgements

The authors acknowledge support from the San Antonio Area Foundation and the Morrison Trust.

Conflict of interest

The authors declare no conflict of interest.

Author details

Philip Serwer*, Elena T. Wright and Cara B. Gonzales
The University of Texas Health Science Center, San Antonio, Texas, USA

*Address all correspondence to: serwer@uthscsa.edu

References

[1] Park GT, Choi KC. Advanced new strategies for metastatic cancer treatment by therapeutic stem cells and oncolytic virotherapy. Oncotarget. 2016;7:58684-58695. DOI: 10.18632/oncotarget.11017

[2] Leaf C. The Truth in Small Doses: Why We're Losing the War on Cancer—And How to Win It. New York: Simon and Schuster; 2014

[3] Hanahan D. Rethinking the war on cancer. The Lancet. 2014;**383**:558-563. DOI: 10.1016/S0140-6736(13)62226-6

[4] Mamdouhi T, Twomey JD, Mc Sweeney KM, Zhang B. Fugitives on the run: Circulating tumor cells (CTCs) in metastatic disease. Cancer Metastasis Reviews. 2019;**38**:297-305. PMID: 27683421

[5] Lambert AW, Pattabiraman DR, Weinberg RA. Emerging biological principles of metastasis. Cell. 2017;**168**:670-691. DOI: 10.1016/j.cell.2016.11.037

[6] Chabner BA, Roberts TG Jr. Timeline: Chemotherapy and the war on cancer. Nature Reviews Cancer. 2005;**5**:65-72. DOI: 10.1038/nrc1529

[7] Nurgali K, Jagoe RT, Abalo R. Editorial: Adverse effects of cancer chemotherapy: Anything new to improve tolerance and reduce sequelae? Frontiers in Pharmacology. 2018;**9**: 245. DOI: 10.3389/fphar.2018.00245

[8] DeVita VT Jr, Chu E. A history of cancer chemotherapy. Cancer Research. 2008;**68**:8643-8653. DOI: 10.1158/0008-5472.CAN-07-6611

[9] Inthagard J, Edwards J, Roseweir AK. Immunotherapy: Enhancing the efficacy of this promising therapeutic in multiple cancers. Clinical Science (London, England). 2019;**133**:181-193. DOI: 10.1042/CS20181003

[10] Christofi T, Baritaki S, Falzone L, Libra M, Zaravinos A. Current perspectives in cancer immunotherapy. Cancers (Basel). 2019;**11**:E1472. DOI: 10.3390/cancers11101472

[11] Marshall HT, Diamgoz MBA. Immuno-oncology: Emerging targets and combination therapies. Frontiers in Oncology. 2018;**8**:315. DOI: 10.3389/fonc.2018.00315

[12] Schaue D, McBride WH. Opportunities and challenges of radiotherapy for treating cancer. Nature Reviews Clinical Oncology. 2015;**12**:527-540. DOI: 10.1038/nrclinonc.2015.120

[13] Chen HHW, Kuo MT. Improving radiotherapy in cancer treatment: Promises and challenges. Oncotarget. 2017;**8**:62742-62758. DOI: 10.18632/oncotarget.18409

[14] Baskar R, Lee KA, Yeo R, Yeoh KW. Cancer and radiation therapy: Current advances and future directions. International Journal of Medical Sciences. 2012;**9**:193-199. DOI: 10.7150/ijms.3635

[15] Zhang Y, Liu Z. Oncolytic virotherapy for malignant tumor: Current clinical status. Current Pharmaceutical Design. 2019. DOI: 10.2174/1381612825666191104090544 [Epub ahead of print]

[16] Lawler SE, Speranza MC, Cho CF, Chiocca EA. Oncolytic viruses in cancer treatment: A review. JAMA Oncology. 2017;**3**:841-849. DOI: 10.1001/jamaoncol.2016.2064

[17] Serwer P, Wright ET. Nanomedicine and phage capsids. Viruses. 2018;**10**: E307. DOI: 10.3390/v10060307

[18] Bloch H. Experimentelle untersuchungen über beziehungen zwischen bakteriophagen und malignen

tumoren. Archiv fur die Gesamte Virusforschung. 1940;**1**:481-496. DOI: 10.1007/BF01240654

[19] Budynek P, Dąbrowska K, Skaradziński G, Górski A. Bacteriophages and cancer. Archives of Microbiology. 2010;**192**:315-320. DOI: 10.1007/s00203-010-0559-7

[20] Kalyane D, Raval N, Maheshwari R, Tambe V, Kalia K, Tekade RK. Employment of enhanced permeability and retention effect (EPR): Nanoparticle-based precision tools for targeting of therapeutic and diagnostic agent in cancer. Materials Science & Engineering. C, Materials for Biological Applications. 2019;**98**:1252-1276. DOI: 10.1016/j.msec.2019.01.066

[21] Fang J, Islam R, Islam W, Yin H, Subr V, Etrych T, et al. Augmentation of EPR E_ect and E_cacy of anticancer nanomedicine by carbon monoxide generating agents. Pharmaceutics. 2019;**11**:343. DOI: 10.3390/pharmaceutics11070343

[22] Anchordoquoy TJ, Barenholz Y, Boraschi D, Chorny M, Decuzzi P, Dobrovolskaia MA, et al. Mechanisms and barriers in cancer nanomedicine: Addressing challenges, looking for solutions. ACS Nano. 2017;**11**:12-18. DOI: 10.1021/acsnano.6b08244

[23] Barenholz Y. Doxil®—The first FDA-approved nano-drug: Lessons learned. Journal of Controlled Release. 2012;**160**:117-134. DOI: 10.1016/j.jconrel.2012.03.020

[24] Silverman L, Barenholz B. In vitro experiments showing enhanced release of doxorubicin from Doxil® in the presence of ammonia may explain drug release at tumor site. Nanomedicine. 2015;**11**:1841-1850. DOI: 10.1016/j.nano.2015.06.007

[25] Nakamura Y, Mochida A, Choyke PL, Kobayashi H. Nanodrug delivery: Is the enhanced permeability and retention effect sufficient for curing cancer? Bioconjugate Chemistry. 2016;**27**:2225-2238. DOI: 10.1021/acs.bioconjchem.6b00437

[26] Gopalakrishna P. Nanomedicines for cancer therapy: An update of FDA approved and those under various stages of development. SOJ Pharmacy and Pharmaceutical Sciences. 2014;**1**:13. DOI: 10.15226/2374-6866/1/2/00109

[27] Abraham SA, Waterhouse DN, Mayer LD, Cullis PR, Madden TD, Bally MB. The liposomal formulation of doxorubicin. Methods in Enzymology. 2005;**39**:71-96. DOI: 10.1016/S0076-6879(05)91004-5

[28] Kanwal U, Irfan Bukari N, Ovais M, Abass N, Hussain K, Raza A. Advances in nano-delivery systems for doxorubicin: An updated insight. Journal of Drug Targeting. 2018;**26**:296-310. DOI: 10.1080/1061186X.2017.1380655

[29] Gabizon A, Martin F. Polyethylene glycol-coated (pegylated) liposomal doxorubicin. Rationale for use in solid tumours. Drugs. 1997;**54**(Suppl. 4):15-21. DOI: 10.2165/00003495-199700544-00005

[30] Serwer P, Wright E, Lee JC. High murine blood persistence of phage T3 and suggested strategy for phage therapy. BMC Research Notes. 2019;**12**:560. DOI: 10.1186/s13104-019-4597-1

[31] Serwer P, Wright ET, Demeler B, Jiang W. States of T3/T7 capsids: Buoyant density centrifugation and cryo-EM. Biophysical Reviews. 2018;**10**:583-596. DOI: 10.1007/s12551-017-0372-5

[32] Serwer P, Wright ET, Chang J, Liu X. Enhancing and initiating phage-based therapies. Bacteriophage. 2014;**4**:e961869. DOI: 10.4161/21597073.2014.961869

[33] Khan SA, Griess GA, Serwer P. Assembly-associated structural changes of bacteriophage T7 capsids. Detection by use of a protein-specific probe. Biophysical Journal. 1992;**63**:1286-1292. DOI: 10.1016/S0006-3495(92)81724-1

[34] Guo F, Liu Z, Fang P-A, Zhang Q, Wright E, Wu W, et al. Capsid expansion mechanism of bacteriophage T7 revealed by multistate atomic models derived from cryo-EM reconstructions. Proceedings of the National Academy of Sciences of the United States of America. 2014;**111**:E4606-E4614. DOI: 10.1073/pnas.1407020111

[35] Serwer P. A metrizamide-impermeable capsid in the DNA packaging pathway of bacteriophage T7. Journal of Molecular Biology. 1980;**138**:65-91. DOI: 10.1016/s0022-2836(80)80005-2

[36] Guo F, Liu Z, Vago F, Ren Y, Wu W, Wright ET, et al. Visualization of uncorrelated tandem symmetry mismatches in the internal genome packaging apparatus of phage T7. Proceedings of the National Academy of Sciences of the United States of America. 2013;**110**:6811-6816. DOI: 10.1073/pnas.1215563110

[37] Watson RA, De La Pena H, Tsakok MT, Joseph J, Stoneham S, Shamash J, et al. Development of a best-practice clinical guideline for the use of bleomycin in the treatment of germ cell tumours in the UK. British Journal of Cancer. 2018;**119**:1044-1051. DOI: 10.1038/s41416-018-0300-x

[38] Motlagh NS, Parvin P, Ghasemi F, Atyabi F. Fluorescence properties of severalchemotherapydrugs:Doxorubicin, paclitaxel and bleomycin. Biomedical Optics Express. 2016;**7**:2400-2406. DOI: 10.1364/BOE.7.002400

[39] Medscape, bleomycin (Rx). 2019. Available from: https://reference.medscape.com/drug/bleomycin-342113

[40] Drugs.com. 2019. Available from: https://www.drugs.com/dosage/bleomycin.html

[41] Dawson NJ. The surface-area/body-weight relationship in mice. Australian Journal of Biological Sciences. 1967;**20**:687-690. ISSN: 0004-9417

[42] Serwer P, Watson RH, Hayes SJ, Allen JL. Comparison of the physical properties and assembly pathways of the related bacteriophages T7, T3 and phi II. Journal of Molecular Biology. 1983;**170**:447-469. DOI: 10.1016/s0022-2836(83)80157-0

[43] Serwer P, Wright ET, Liu Z, Jiang W. Length quantization of DNA partially expelled from heads of a bacteriophage T3 mutant. Virology. 2014;**456-457**:157-170. DOI: 10.1016/j.virol.2014.03.016

Chapter 3

New Generation Peptide-Based Vaccine Prototype

Öznur Özge Özcan, Mesut Karahan,
Palanirajan Vijayaraj Kumar, Shen Leng Tan and Yi Na Tee

Abstract

Synthetic peptide-based vaccine prototypes are the future potential vaccination. Antigens, which belong to minimal microbial component and produce antibodies such as peptides and polysaccharides, can promote long-term protection against pathogens that can cause infectious diseases. Production of peptides becomes simple with solid phase peptide synthesis and microwave-assisted solid phase peptide synthesis using automatic synthesizers. The use of synthetic peptides was approved by the health authorities for vaccine design. Peptides are themselves very weak immunogens and need adjuvants to provide an effective autoimmune response. For this reason, peptide antigens are conjugated with biopolymers and loaded with nanoparticles. The toxicity of vaccine prototypes is evaluated in cell culture, and non-toxic prototypes are selected for vaccinating experimental animals. The most effective peptide-based vaccine prototype is determined as the one with the highest antibody level. The goal of this book chapter is to illustrate the use of peptides vaccine systems and present their opportunities with their future development.

Keywords: biopolymers, nanoparticle systems, solid phase peptide synthesis (SPPS), synthetic peptide, peptide vaccine prototype

1. Introduction

The goal of this chapter is to review the importance of synthetic peptide-based vaccination, providing a brief knowledge about their new generation prototypes. In the first stage, relation to immunity and peptide vaccine with importance of using biopolymers was given under the title of solid-phase peptide synthesis including microwave system. After that, this review was focused on the established methods for peptide loaded nanoparticles or conjugated biopolymers preparation of peptide-based vaccine prototypes and nanotechnological particles as delivery system with touching on different methods. In addition, the impact of Contemporary Advancements in Peptide Based Vaccine like Liposome Based Subunit Vaccines was explained. In the last part, peptide-based vaccine prototypes studies *in vivo and in vitro* were given with their future perspective and development.

2. Peptide vaccines prototype and immunity

All vaccines generally are developed by using live or attenuated microorganisms. However, the use of whole microorganisms, their components or the biological process for vaccine production has many weaknesses and a variety of approaches for synthetic peptide vaccination remain under investigation for the infectious diseases [1]. Peptides play an important role in a biological process, including the stimulate the immune response [2].

Peptide-based vaccination is an immunotherapy where a peptide is applied often with the use of an immunoadjuvant (nanoparticle or biopolymers) to stimulate T-cell and sometimes B-cell immunity. Peptide-based vaccinations are present in major histocompatibility complexes (MHC) the ultimate target for T cells in infection recognition and infection immune responses [3, 4]. Sometimes peptide-based vaccines play a role to stimulate innate and adaptive immunity both (**Figure 1**) and peptides are immunogen components of peptide-based vaccine and memory responses of peptide is weak in immune responses [1] without the biopolymer or nanoparticle system.

When producing a new generation of synthetic peptide vaccines, components of the pathogenic pathogen of interest are generally used. These components are linear peptides and produced by solid phase peptide synthesis (SPPS) method with high efficiency and purity [5–9]. When peptides are used in combination with a vaccine system, if they are used without a drug delivery system, there are risks of degradation by protease enzymes that break down proteins and phagocytosis by immune system cells such as antibodies [10]. In addition, drug delivery systems should be preferred as nanoparticles and biopolymers. A higher immune response

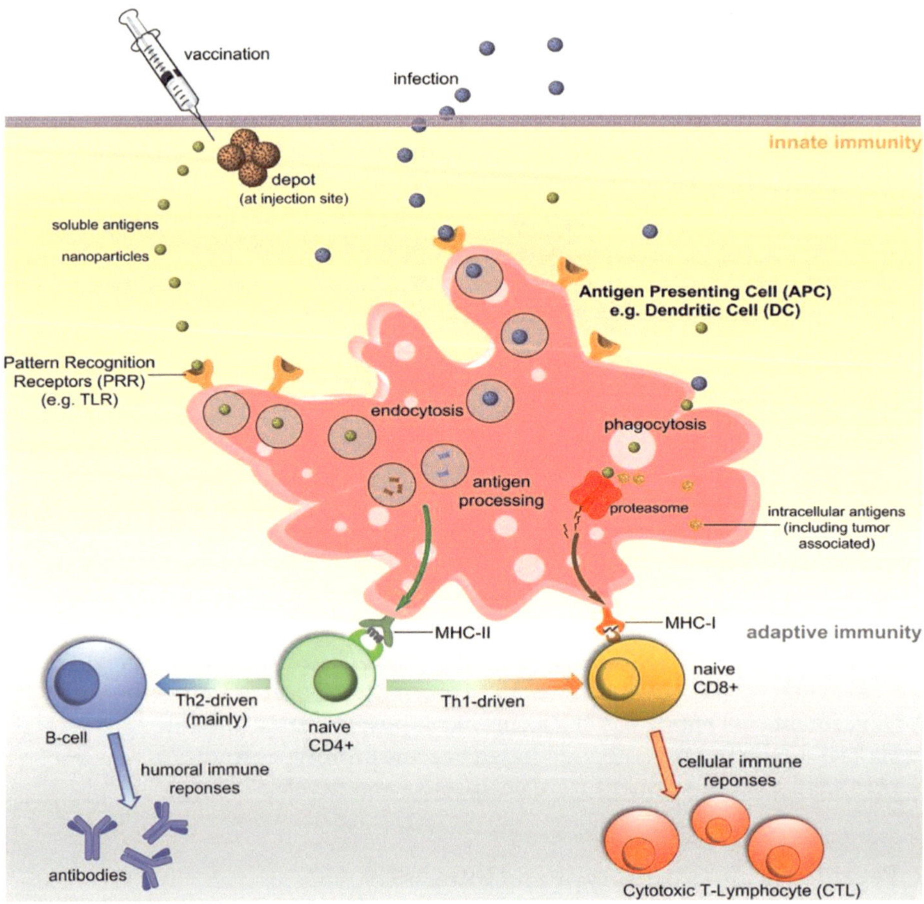

Figure 1.
Cellular representation of immune cells after vaccination [1].

and to protect peptides from harmful effects of degrading enzymes and aggressive antibodies, it is generally necessary to use nanosystems such as protein, biopolymer conjugations or nanoparticles (NPs). Peptide delivery systems based on nanoparticles are developing more and more for the development of peptide-based vaccines. Especially, biodegradable polymers offer very popular and patented vaccines [11]. For instance, poly(amino acid) and polylactic acid (PLA) are used for NPs and when antigenic peptides are encapsulated by them in order to vaccinate mice can provide significantly higher levels of total the antibodies like immunoglobulins (IgGs); IgG, IgG1, IgG2. This means they can able to stimulate humoral immune responses and also CD4+ and CD8+, T and B cell activation and for the cellular immune responses; interferon γ (IFNγ) which induce Ig class switching to IgG2a [12–14]. In another study, Murine model was used in an immunization study. An antigen of Hepatitis B disease loaded on Poly (lactic-co-glycolic acid) (PLGA) NPs (300 nm) provided better immune responses compared to the antigen alone. Immunization with PLA NPs (200–600 nm) can also provide higher levels of IFNγ production related to a Th1 response. In contrast, immunization of PLA microparticles (2–8 μm) promoted IL-4 secretion due to Th2 response [15]. Both PLGA NPs and liposomes are phagocytosed efficiently by cells to localize intracellular localizations and produce an immune response [16, 17]. Carbon NPs are promising in oral vaccine administration for the use of synthetic peptides [18].

Different approaches are available to develop synthetic peptide-based vaccines, using metal ions in combination with peptide sequences. In particular, the investigation of the complex formation biopolymer by peptide in the presence of metal ions contributes greatly to the technological development of peptide-based vaccine prototypes [19]. The contact of the peptides with the polyelectrolyte (PE) is found at the interface. Solubility of polyplexes and complexes with NPs and peptides; it depends on the structure of the peptides (such as hydrophilic and lipophilic) and correlates with the isoelectric points in this system. Metal ions such as copper (Cu^{+2}) generally promote two effects: (1) conjugation of polyelectrolyte to peptide molecules and (2) aggregation of polyplex particles in the intermolecular region. Some of these polyplexes exhibit strong immunogenicity and provide a high level of immunological protection for peptide vaccine prototypes, making them more efficient, but the solubility, composition and stability of these polycomplexes depend on pH, metal/PE and protein/PE ratios. These systems are based on conjugation of PE and antigen molecules with covalent bonds to NPs or biopolymers, which induce an immune response to the immunizing agent. The hydrophobic interactions in such a complex create an adjuvant effect for prototyping technology in vaccination. [19–22]. In the studies on the development of peptide vaccine prototypes previously made by our study group, it was observed that the purification of characterization of binding of synthetic peptides to various adjuvants and subsequent high immune response was obtained in BALB/c mice from experimental animals [23].

3. Solid-phase peptide synthesis (SPPS)

A historical overview of peptide chemistry from T. Curtius (who achieved the first synthesis of peptide in 1882) and Fischer (who synthesized the first dipeptide in 1901) to M. Bergmann and L. Zervas is first in presenting the Solid-Phase peptide synthesis. Next, the fundamentals of peptide synthesis with a focus on SPPS by R. B. Merrifeld are described. Although the peptides can be synthesized in three methods: in a solution medium, on a solid support, or as a combination of the solid and the solution synthesis, this chapter emphasizes an overview of peptide synthesis giving importance on SPPS. Currently, most of the peptides for research, vaccination or therapeutic drugs for cancer and brain diseases are synthesized by SPPS methods. Successful peptide synthesis depends on the appropriate selection of suitable resins, linkers, amino acid derivatives and coupling reagents, as well as the side chain (de) protection and cleavage conditions, and the correct synthesis of the assay. In the SPPS method, the solid support is attached at the end of the first amino acid-COOH at the carboxyl end a polymeric support insoluble in the newly formed peptide chain is referred to as resin. A covalent binding step that binds the resin is important for the reaction [24]. The peptides may be gradually joined between the C and N terminus using N-protected amino acids. The Nα protecting group (Boc) is unstable in the presence of intermediate acid (trifluoroacetic acid; TFA), the side chain protecting benzyl (Bzl) based groups and the peptide/resin linkage are stable in the presence of intermediate acid and are variable in the presence of strong acid (HF). Fmoc group is important for solid-phase applications. Fmoc-based strategies are also available, and hydroxymethylphenoxy-based binders are used to add peptide to the resin with t-butyl (tBu) based side chain protection [25]. The solid phase peptide synthesis method consists of three basic steps. According to this, deprotection of the carboxyl group activation and peptide bond formation (Coupling). Following this procedure, the final deprotection of the last added amino acid is removed and the N- terminal is released. Cleavage and deprotection of the resin-bound peptide from the solid support [26].

The stepwise representation of solid phase peptide synthesis is illustrated in **Figure 2**. The starting amino acid masked by a non-persistent protecting group at the N-α terminus is loaded from the C-terminus to the resin. A semi-permanent protection group can also be used to mask the side chain if necessary (**Figure 3**, Step 1). The synthesis of the peptide, repeated deprotection of the N-α-transient protecting group, and binding of the next protected amino acid (**Figure 3**, Step 3).

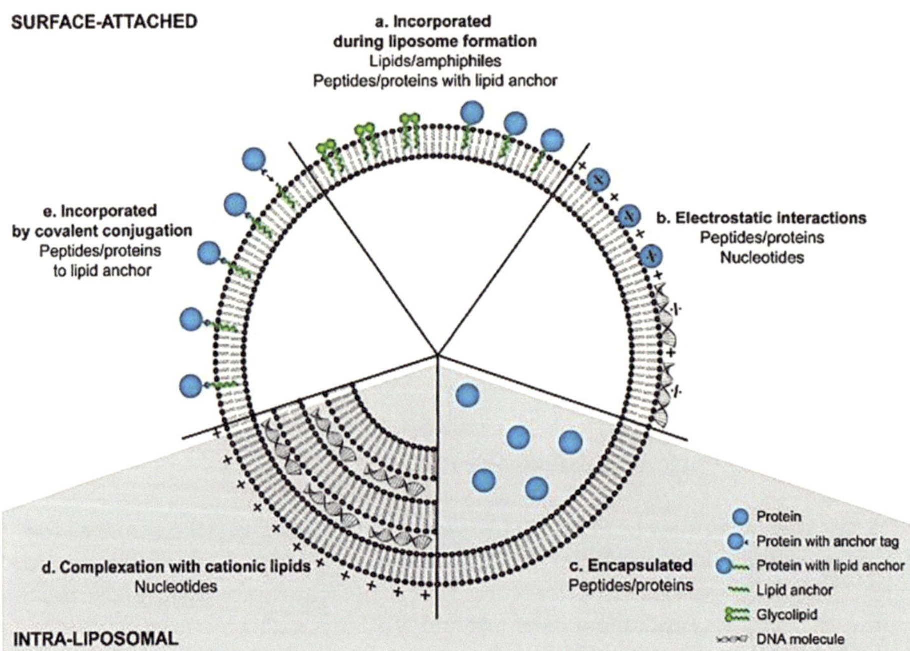

Figure 2.
The antigens and immunomodulators that can be used for inclusion in liposomes; it is shown in different strategies depending on the target and structure of the molecule [28].

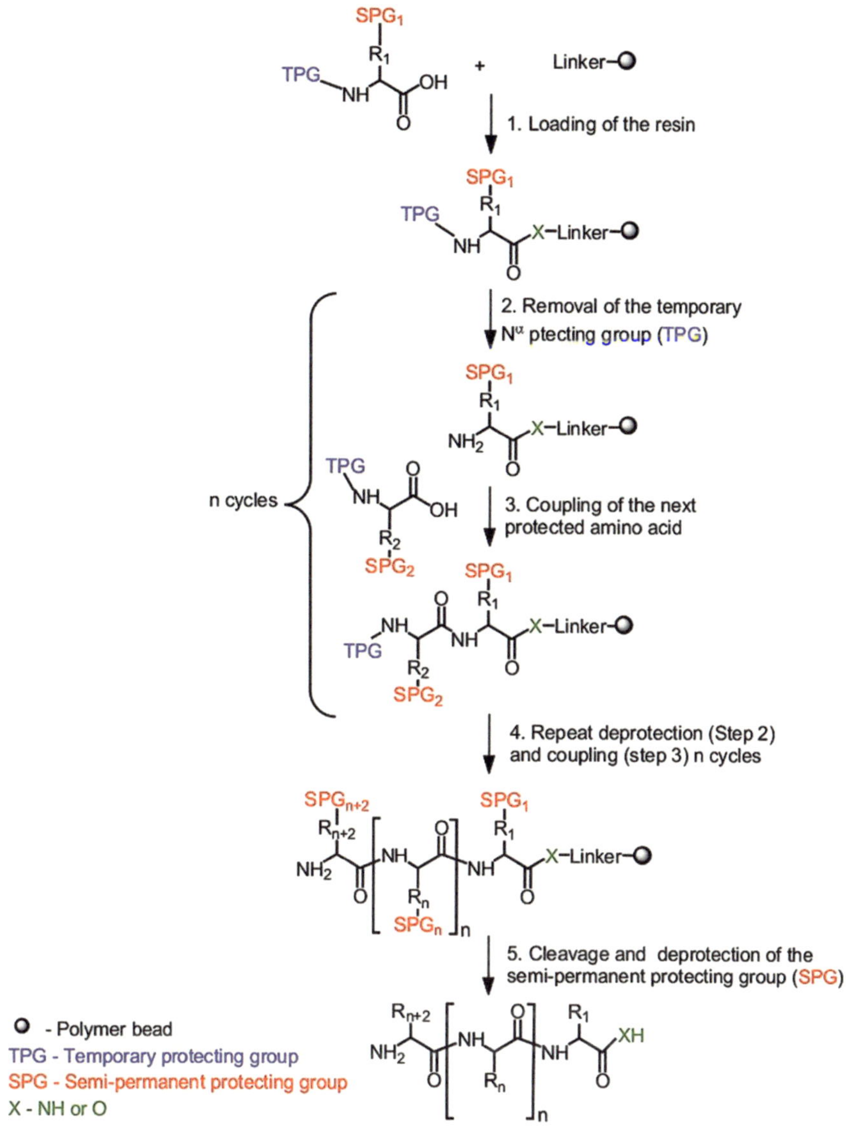

Figure 3.
Stepwise representation of solid phase peptide synthesis [25].

Synthetic cycle	Reagents	Time & conditions
Deprotections	Trifluoroacetic acid (Boc) and 20% piperidine in DMF (Fmoc)	1–5 min (70°C Fmoc and Boc)
Couplings	Amino acids, HBTU/HATU/HOBt/HOAt/DIC, DIPEA	5–15 min 50–70°C

Table 1.
Synthetic cycle and important reagents with time and conditions in microwave-assisted SPPS [27].

After the last amino acid is loaded (**Figure 3**, Step 4), the peptide is separated from the resin support and the Fmoc or Boc groups are removed [25].

The development of microwave-assisted solid-phase peptide synthesis has been developed by the synthesis of linear and complex peptide sequences and long peptide sequences in a shorter time and high throughput. Time and temperature conditions, reagents and synthesis cycle for microwave assisted SPPS with using Boc or Fmoc are represented below in **Table 1**. The disadvantage of this technique could be the cost of resin (the binding procedure of the first binding amino acid to the resin in peptide synthesis can requires different and complex processes) and equipment [27, 29].

4. Peptide-based vaccine for nanotechnological prototypes

Synthetic peptides alone are not sufficient to develop vaccine prototypes because they cannot stimulate the cellular and humoral protection system sufficiently. So, different adjuvant systems are formed by conjugating the peptides biopolymers or loading them into NPs, resulting in a high immune response [30]. The use of peptide-polymer complexes and peptide loaded nanoparticles are the best way for the developing the peptide-based vaccine prototypes.

4.1 Type of nanoparticles

The binding of the antigenic peptides with the water-soluble polymer has multiple effects. Some of those are as follows:

- to provide modification of peptides,
- to increase the water solubility of those with hydrophobic properties,
- to raise regional impact,
- to increase immunogenic effects and immunoreactivity, and
- to be more effective in the living organism [31].

NPs are spherical polymeric carriers. The particles that are below 1000 nanometers (nm) are called nanoparticles. These particles with superior properties are used in many fields such as electricity, electronics, biotechnology, automotive, medical. NPs are morphologically and physicochemically influenced by the physical and chemical properties of the starting material used. The nanoparticles used as polymeric carriers are solid colloidal structure. The active substance can be encapsulated, absorbed or dissolved in the particle. Polysaccharides, polyanhydride, polycaprolactone, polyacrylic acid and polylactic-co-glycolic acid is also used for producing an effective nanoparticles and produce a co-polymer system such as poly(ethylene glycol) (PEG)-Nps and poly(ethylene glycol)-poly(ε-caprolactone) copolymers (PEG-PCL) copolymers. The copolymers of N-vinyl-2-pyrrolidone with acrylic acid (P(VP-co-AA)), PLGA, NPs loaded with the antigenic peptide can be used for future vaccine prototypes [32]. PLGA NPs is approved by U.S. Food and Drug Administration (FDA) and using for peptide carrier in vivo because of strong immune response [33].

Polymeric NPs are used for therapeutic applications and some of popular NPs are as follows:

1. Pluronics®
2. PEG–PLA

3. PEG-PCL

4. PEG-Lipid

5. PEG-PLGA

6. PEG-poly (amino acids)

7. Stimuli-sensitive polymeric micelles

8. Endogenous stimuli-sensitive polymeric micelles

9. pH-sensitive polymeric micelles

10. Reduction sensitive polymeric micelles

11. Thermo-sensitive polymeric micelles

12. Exogenous stimuli-sensitive polymeric micelles

13. Light-sensitive polymeric micelles

14. Magnetic field-sensitive polymeric micelles

15. Ultra-sound sensitive polymeric micelles

16. Margination of micro/NPs: Requirement for optimum drug delivery

5. Established methods for peptide loaded NPs or conjugated biopolymers preparation

We have mentioned that the peptides alone cannot produce an adequate immune response and also have poor stability with the internalization problem while crossing cell membranes. To solve all these limitations, peptides are loaded nanoparticle systems or conjugated biopolymers. Biopolymers are generally nontoxic products are generally preferred for producing continuously release systems with long term effect [34]. Here, the applicable and most common strategies for the synthesis of peptide-based NPs and encapsulation or conjugated methods of biopolymers are shown.

5.1 Emulsification-solvent evaporation method

The emulsion solvent evaporation technique is known as the most successful and useful method in the preparation of peptide loaded NPs and this technique is studied under two groups as single and double emulsion solvent evaporation methods [35].

5.2 Conjugation methods

Conjugation is a technique for achieving peptide and biopolymer complexes. The covalently linked peptide biopolymer conjugates can be linked using the water-soluble carbodiimide method as a cross-linker and synthesis with microwave energy methods [36]. Peptides conjugates biopolymers can be synthesized in organic media using microwave energy. Also, there are another methods, including complex formation of biopolymers and peptides and electrostatic complex formation and metal coordination via ion coordination [35]. Specific antibody titers were observed in mouse experiments against peptides containing polymeric conjugates and complexes. The molecular weights of these conjugates are also very important. Biopolymer conjugation is crucial to obtain a high immune response to antigens at low molecular weights [37–39].

5.3 Nanoprecipitation

Nanoprecipitation is the most strategic method for the preparing of vaccine prototypes. Reducing the pH is very important to stabilization of system. Also, salt concentration under the solubility conditions is another important thing for the encapsulation method [70]. If the experiments cannot move on then adding a non-solvent phase in the quality of the solvent technique in which the parent compound of the NPs is dissolved can help [40]. Nanoprecipitation is frequently used in encapsulation of peptides. A pH-controlled precipitation rather than a non-solvent precipitation is a more preferred approach for passing the polymer to a non-dissolved phase with a simple pH change in the medium. For NPs or biopolymers prepared by nanoprecipitation, these solvents are known as the organic phase of acetone and ethanol [41].

5.4 Encapsulation of peptide

Encapsulation is carried out simultaneously by synthesizing NPs and biopolymers in all of the methods mentioned the encapsulation method for peptides should be selected based on the hydrophobic or hydrophilic facilities of peptide. Using of peptide encapsulation is important because of

- peptide release controlling,
- modeling of targeted delivery systems,
- mask unfavorable organoleptic properties (taste, odor, color),
- protection of peptide from immune attacks and enzyme degradation,
- insurance of bioconjugate molecules stability,
- decrease toxicity, and
- design of new dosage forms [42].

5.5 Peptide characterization

After purification of the peptides, they are commonly characterized by liquid chromatography-electrospray ionization-mass spectrometry (LS-ESI-MS), fluorescence spectroscopy and possible three-dimensional structures of the synthetic peptide (PEP-FOLD) server. It has validation since the chromatographic method has positive properties in terms of linearity, accuracy, precision and repeatability. Synthetic peptide vaccines are immunogens that can be used when creating vaccine prototypes, especially because of their lipophilic structure (which also allows cell permeability to pass easily) [43]. Methods such as Fourier transform infrared (FT-IR) and nuclear magnetic resonance (1H- and 31P-NMR) are frequently used to visualize the physical structure of the copolymer and peptide biopolymer conjugates and to perform characterization studies. The conjugation of molecular weights is measured via size-exclusion chromatography (SEC) [44].

5.6 Characterization of peptide vaccine prototype

Ultraviolet (UV) and FT-IR Spectrophotometers and ZetaSizer are used for studying the nanoparticles and Scanning Electron Microscope (SEM) is used for morphological examination of Polymers or bioconjugate [45].

5.7 Toxicity studies

Peptide-based vaccine prototypes need to be tested in a cell culture medium to be feasible because they may have physiological, biological and chemical effects, causing cytotoxicity. The method used to investigate the cytotoxic profiles of peptide-based vaccines is also called *in vitro* cytotoxicity assays or cell culture-based measurement methods [46, 47]. Tetrazolium salts are compounds used in cell lines to measure the metabolic pathways of cells of microbial origin. Tetrazolium salts are the heterocyclic organic structure of these compounds and their reduction to colorless or weak colored aqueous solutions known as formazans has been the basis of their use as vital dyes in redox chemistry, biological and chemical applications [46, 47]. The tetrazolium ring can only be broken by active mitochondria, so viable cells and dead cells can be distinguished by discoloration. The fact that this change can be made only by living cells in vitro has made tetrazolium compounds a highly biologically important to measure toxicity of peptide-based vaccine formulas. The mechanism of toxicity assays, such as 3-(4,5-dimethylthiazolyl)-2,5 diphenyltetrazolium bromide (MTT) [19], 3-(4,5-dimethylthiazol-2yl)-5-(3-carboxymethoxyphenyl)-2-(4-sulfophenyl)-2H-tetrazolium (MTS), 2,3-bis(2-methoxy-4-nitro-5-sulfophenyl)-5[(phenylamino)carbonyl]-2H-tetrazolium hydroxide (XTT), sodium 5-(2,4-disulfophenyl)-2-(4-iodophenyl)-3-(4-nitrophenyl)-2Htetrazolium inner salt (WST), 5-methyl-phenazinium methyl sulfate(PMS), 5-[3-(carboxymethoxy)phenyl]-3-(4,5-dimethyl-2-thiazolyl)-2-(4- sulfophenyl)-2H-tetrazolium inner salt; (MTS) are used [49, 50] and in our studies, we generally use MTT analysis. For example, our technological vaccine prototype example is Zika peptide loaded PLGA nanoparticles which were determined on ECV304 human epithelial cells via MTT assay, which is the cytotoxicity test, was performed to determine the cytotoxic effects of the peptide, peptide loaded NPs [45]. The importance of toxicity studies is to determine the non-toxic vaccine prototype and to switch to in vivo animal studies.

5.8 Contemporary advancements in peptide based vaccine

5.8.1 Liposome based subunit vaccine

Live attenuated vaccine is highly immunogenic and considered as well-tolerant for healthy individuals. However, live attenuated vaccine should not be administered to immunocompromised individual as it would cause systemic infection. An alternative vaccine technology, subunit vaccine, is safer and more suitable for immunocompromised individual. It uses fragment of a pathogen (antigen) to trigger an immune response and stimulate immunity against the pathogen. However,

Figure 4.
General features of virus-like particle [48].

subunit vaccine has low immunogenicity and often combined with adjuvants to induce protective immunity. The adjuvants are capable to enhance vaccine effectiveness and stimulate immune responses [28].

Human alphaherpesvirus 3 (HHV-3) is known as Varicella-zoster virus (VZV), which is the causative agent of varicella (chicken pox) and herpes zoster (shingles). In 1978, the first commercial Varicella-zoster immune globulin, Vzig™ (Massachusetts Public Health Biologic Laboratories, Boston, Massachusetts) became available. However, the supply was removed from the U.S. market in 2006. The alternative preparation, Varizig® (Cangene Corporation, Winnipeg, Canada) was licensed by FDA in 2012 and has shown to be comparable to Vzig™ [Saol Therapeutics Inc. 2012]. Varizig® is supplied as a sterile solution containing human Varicella-zoster immune globulin (IgG) which showed for post-exposure prophylaxis in high-risk patients [51].

Zostavax® is the vaccine licensed for herpes zoster prevention in individuals above the age of 50. It is a lyophilized preparation which is given as subcutaneous injection [52]. A non-replicating liposome-based subunit vaccine (HZ/su) is the new development for zoster prevention. The HZ/su is a non-live recombinant VZV glycoproteins E with the adjuvant ASO1B [53]. A randomized placebo-controlled study has shown that HU/su poses age-independent defense against HZ and has better efficacy compared to Zostavax® in reducing the risk of HZ for immunocompromised adults with the age above 50. Unlike HZ/su, Zostavax® lose efficacy as age increase [54]. HZ/su vaccine is not yet approved by FDA.

Liposomes are nano-carriers and they are useful in delivering vaccine antigen by forming liposome-based vaccine delivery systems. It is advantageous over other carriers due to its biocompatibility, non-toxic and biodegradable features [55]. Besides, liposomes can be customized to achieve desired immune profiles by optimizing their composition, antigen-loading strategies and the use of adjuvants system [28, 56].

5.8.2 Virus-like particle

Eculizumab (Soliris®) is a humanized monoclonal antibody (mAb) which function as a terminal complement inhibitor [57]. It was the first therapeutic agent approved by the FDA for atypical hemolytic uremic syndrome (aHUS) and paroxysmal nocturnal hemoglobinuria (PNH)-associated with thrombotic microangiopathy (TMA) in 2007 and 2011 respectively [58]. Soliris® (Alexion Pharmaceuticals, New Haven, Connecticut, USA) is in the form of sterile solution for i.v. injection. Eculizumab increases the patient's susceptibility to meningococcal infection (Neisseria meningitidis), all patients must be vaccinated against meningococcal infections prior to or at the time of initiating Eculizumab.

Virus-like particles (VLPs) is one of the alternative types of nanoparticles delivery system [59]. A recent research has shown that the development of autologous C5 vaccine in nanoparticle form is able to elicit strong humoral responses [60]. A peptide epitope (PADRE peptide) in the C5 vaccine is used to create a recombinant virus-like particles (VLPs). It showed a reduction in hemolytic activity and protect the mice from complement-mediated intravascular hemolysis [60]. Based on the study's result, it is showed that the recombinant VLPs could be used as an alternative or supplement for Eculizumab.

VLPs is known as an emerging class of targeted delivery vehicles with potential of overcoming the limitations of other nanoparticles [48]. VLPs is a potential delivery system due to their immunogenic nature, well-defined structure, ability to present a wide variety of potential epitopes, and ease of production [60]. They lack natural genome thus it is non-infectious. Besides, it can turn as self-adjuvant which is proficient in breaking the immune tolerance.

One of the limitations of VLPs are phagocyte-mediated clearance [59]. Besides, a recent study showed that ellipsoid nanoparticles can extravasate from the blood vessel more effectively than spherical nanoparticles. Meanwhile, the ellipsoid shape is possible for conventional polymeric NPs, but is not feasible for icosahedral VLPs. However, this limitation can be overcome through the modification of VLP surface by adding a variety of useful ligands [61]. VLPs may be able to efficiently extravasate from the vasculature of the blood vessels by showing multiple ligands with high affinity for the tight connections between endothelial cells [59].

6. Importance *in vitro* and *in vivo* experiments using peptide-based vaccine prototypes

After forming a synthetic vaccine prototype, the cytotoxicity of the bioconjugate of the peptides and biopolymers is first determined (generally we use MTT analysis). After the apoptotic effect of the prototype on living cells is measured by flow cytometric detection, the vaccine prototype with the most viable cell number should be selected for further study [62]. After all these methods, immunization is the next step. We immunize BALB/c mice with each one of the peptides biopolymer conjugates or peptides loaded nanoparticles following conventional immunization protocol. The goal is to identify the most antigenic vaccine prototype. The antibodies are measured in blood (for humeral response such as; T and B lymphocytes, IgGs) or splenic (cellular response like ILs and IFNs) samples from the immunized BALB/c mice via the indirect enzyme-linked immunosorbent assay (ELISA) to determinate the highest antibody level. Thus, the most suitable peptide-based vaccine prototypes will be identified for future clinical phase studies. In brief, cell culture and toxicity studies are important before the analyses the effect of vaccine prototype in vivo [62].

7. Current situation and future perspective

Peptides can affect important brain regions that are essential for life-sustaining functions. There are studies about Peptide drug and Peptide-Polymer Vaccines and drugs using for brain disorders. In this future perspective we will explain the using of peptides in common brain disorders such as Alzheimer Disease, Parkinson Disease (PD), Multiple Sclerosis. In a study using an interference-inducing peptide (TAT-DATNT) to elute a protein complex consisting of interaction between DAT and the dopamine D2 receptor (D2R), it was determined that locomotor behaviors were induced in Sprague–Dawley (SD) rats. This peptide can provide potential therapy for regulating the activity of DAT and dopaminergic neurotransmission of Attention Hyperactivity Deficit Disorder (ADHD) therapies [63]. Alpha Synuclein (A-syn) aggregate is very important for the PD. Against of this aggregate, an immunogenic peptide the sequence of CGGVDPDN [64] is developed with solid phase peptide synthesis method as a vaccine in PD. Peptides can also be neuroprotection for the PD. TFP5 peptide, FITCGGGKEAFWDRCLSVINLMSSKMLQINAYARAARRAARR; TP5 peptide, KEAFWDRCLSVINLMSSKMLQINAYARAARRAARR; SCP peptide, FITCGGGGGGFWDRCLSGKGKMSSKGGGINAYARAARRAARR are reduction in neuroinflammation and apoptosis. **MPTP (1-methyl-4-phenyl-1,2,3,6-tetrahydropyridine)** is a neurotoxin drug of MPP+, which causes permanent symptoms of PD by destroying dopaminergic neurons in the substantia nigra of the brain for mouse modeling [65]. PD01A is a Phase 1 epitope vaccine in experimental — a synthetic a-syn mimicking peptide-polymer based on a-syn aggregate by inducing an immune response that generates antibodies specifically against it [66]. In general, vaccines are developed according to the T cell response, but a peptide epitope with a three-celled peptide epitope with a three-cell universal peptide such as Syn85-99 (AGSIAAATGFVKKDD), α-Syn109-126 (QEGILEDMPVDPDNEAYE), α-Syn126-140 (EMPSEEGYQDYEPEA) and P30 (*FNNFTVSFWLRVPKVSASHLE*) epitope vaccines comprising three peptide-based epitope vaccines comprising different α-Syn peptides, but consisting of different B cell epitopes as follows, are noted for their high immunogenicity [67]. So peptides can have different immune cell response in brain disorders. Peptides can also protect cell biological agents such as microtubules activity. This is very important for cell stability while the diseases are seen in the cells. The dysregulation of ADNP / ADNP2 expression in the relevant brain tissue and animal model may improve the prognosis of schizophrenia because these genes are responsible for the regulation of interacting microtubules. The microtubule-interacting drug candidate, NAP (davunetide) is a small peptide and belong to activity-dependent neuroprotective protein (ADNP) which contains a small peptide motif, *NAPVSIPQ* sequence that provides potent neuroprotection for tau pathology, neuronal cell death as well as social and cognitive dysfunctions [68]. Especially in Amyloid beta pathology, *DAEFRHDSGY* peptide, Wang et al. synthesized peptide immunogens, A1-14 peptide immunogens for *UBITh*® AD immunotherapeutic vaccine by using automated SPPS for the Alzheimer Disease.

Wang's group has developed a synthetic peptide vaccine prototype for the prevention and treatment of AD and is conducting *phase II* clinical trials. The occurrence of Alzheimer's disease constitutes a strong immune response to Amyloid Beta (Ab) Plaques; *UB-311* was constructed with two synthetic Ab1-14 targeting peptides (B cell epitopes), each bound to different helper T cell peptide epitopes and formulated in a Th2 delivery system [69].

8. Conclusion

Consequently, this chapter provides a brief manual for anyone in the fields of solid-phase peptide synthesis, peptide vaccines, Nanotechnological importance for effective vaccine prototypes, and their future perspective for other diseases such as brain disorders.

Conflict of interest

The authors declare no conflict of interest.

Author details

Öznur Özge Özcan[1], Mesut Karahan[1*], Palanirajan Vijayaraj Kumar[2], Shen Leng Tan[2] and Yi Na Tee[2]

1 Üsküdar University, Istanbul, Turkey

2 UCSI University, Kuala Lumpur, Malaysia

*Address all correspondence to: mesut.karahan@uskudar.edu.tr

References

[1] Skwarczynski M, Toth I. Peptide-based synthetic vaccines. Chemical Science. 2016;**7**(2):842-854. DOI: 10.1039/c5sc03892h

[2] Jaradat DMM. Thirteen decades of peptide synthesis: Key developments in solid phase peptide synthesis and amide bond formation utilized in peptide ligation. Amino Acids. 2017;**50**(1):39-68. DOI: 10.1007/s00726-017-2516-0

[3] Maryanski JL, Pala P, Corradin G, Jordan BR, Cerottini JC. H-2-restricted cytolytic T cells specifi c for HLA can recognize a synthetic HLA peptide. Nature. 1986;**324**:578-579. DOI: 10.1038/324578a0

[4] Townsend AR, Gotch FM, Davey J. Cytotoxic T cells recognize fragments of the influenza nucleoprotein. Cell. 1985;**42**:457-467. DOI: 10.1016/0092-8674(85)90103-5

[5] Merrifield RB. Solid phase peptide synthesis. I. The synthesis of a tetrapeptide. Journal of the American Chemical Society. 1963;**85**:2149-2154. DOI: 10.1021/ja01056a056

[6] Nava-Parada P, Forni G, Knutson KL, Pease LR, Celis E. Peptide vaccine given with a toll-like receptor agonist is effective for the treatment and prevention of spontaneous breast tumors. Cancer Research. 2007;**67**: 1326-1334. DOI: 10.1158/0008-5472. CAN-06-3290

[7] Oka Y, Tsuboi A, Oji Y, Kawase I, Sugiyama H. WT1 peptide vaccine for the treatment of cancer. Current Opinion in Immunology. 2008;**20**:211-220. DOI: 10.1016/j.coi.2008.04.009

[8] Sciutto E et al. Improvement of the synthetic tri-peptide vaccine (S3Pvac) against porcine Taenia solium cysticercosis in search of a more effective. Inexpensive manageable vaccine. Vaccine. 2007;**25**:1368-1378. DOI: 10.1016/j.vaccine.2006.10.018

[9] Wang XJ et al. Preparation of a peptide vaccine against GnRH by a bioprocess system based on asparaginase. Vaccine. 2010;**28**:4984-4988. DOI: 10.1016/j. vaccine.2010.05.026

[10] Dai C, Wang B, Zhao H. Microencapsulation peptide and protein drugs delivery system. Colloids and Surfaces. B, Biointerfaces. 2005;**41**(2):117-120. DOI: 10.1016/j. colsurfb.2004.10.032

[11] Manish M, Rahi A, Kaur M, Bhatnagar R, Singh S. A single-dose PLGA encapsulated protective antigen domain 4 nanoformulation protects mice against bacillus anthracis spore challenge. PLoS One. 2013;**8**(4):e61885. DOI: 10.1371/journal.pone.0061885

[12] Uto T, Wang X, Sato K, Haraguchi M, Akagi T, Akashi M, et al. Targeting of antigen to dendritic cells with poly(gammaglutamic acid) nanoparticles induces antigen-specific humoral and cellular immunity. Journal of Immunology. 2007;**178**:2979-2986. DOI: 10.4049/jimmunol.178.5.2979

[13] Uto T, Akagi T, Hamasaki T, Akashi M, Baba M. Modulation of innate andadap-tiveimmunitybybiodegradable nanoparticles. Immunology Letters. 2009;**125**:46-52. DOI: 10.1016/j. imlet.2009.05.008

[14] Mohr E, Cunningham AF, Toellner K-M, Bobat S, Coughlan RE, Bird RA, et al. IFN-γ produced by CD8 T cells induces T-bet– Dependent and – independent class switching in B cells in responses to alum-precipitated protein vaccine. Proceedings of the National Academy of Sciences of the United

States of America. 2010;**107**:17292-17297. DOI: 10.1073/pnas.1004879107

[15] Gutierro I, Hernández RM, Igartua M, Gascón AR, Pedraz JL. Size dependent immune response after subcutaneous, oral and intranasal administration of BSA loaded nanospheres. Vaccine. 2002;**21**:67-77. DOI: 10.1016/S0264-410X(02)00435-8

[16] Copland MJ, Baird MA, Rades T, McKenzie JL, Becker B, Reck F, et al. Liposomal delivery of antigen to human dendritic cells. Vaccine. 2003;**21**:883-890. DOI: 10.1016/S0264-410X(02)00536-4

[17] Elamanchili P, Diwan M, Cao M, Samuel J. Characterization of poly(d, llactic-co-glycolic acid) based nanoparticulate system for enhanced delivery of antigens to dendritic cells. Vaccine. 2004;**22**:2406-2412. DOI: 10.1016/j.vaccine.2003.12.032

[18] Wang T, Zou M, Jiang H, Ji Z, Gao P, Cheng G. Synthesis of a novel kind of carbon nanoparticle with large mesopores and macropores and its application as an oral vaccine adjuvant. European Journal of Pharmaceutical Sciences. 2011;**44**:653-659. DOI: 10.1016/j.ejps.2011.10.012

[19] Karahan M, Mustafaeva Z, Özeroğlu C. The formation of polycomplexes of poly(methyl vinyl ether-Co-maleic anhydride) and bovine serum albumin in the presence of copper ions. Polish Journal of Chemical Technology. 2014;**16**(3):97-105. DOI: 10.2478/pjct-2014-0058

[20] Karahan M, Mustafaeva Z, Ozer H. Polysaccharide-protein covalent conjugates and ternary metal complexes. Asian Journal of Chemistry. 2007;**19**:1837-1845

[21] Karahan M, Mustafaeva Z, Özeroğlu C. Investigation of ternary complex formations of polyacrylic acid with bovine serum albumin in the presence of metal ions by fluorescence and dynamic light scattering measurements. The Protein Journal. 2010;**29**:336-342. DOI: 10.1007/s10930-010-9257-1

[22] Karahan M, Tuğlu S, Mustafaeva Z. Synthesis of microwave-assisted poly(methyl vinyl ether-co-maleic anhydride)-bovine serum albumin bioconjugates. Artificial Cells, Blood Substitutes, and Biotechnology. 2012;**40**(6):363-368. DOI: 10.3109/10731199.2012.678942

[23] Karahan M. Antigenic peptide synthesis and characterization of Q fever disease. Afyon Kocatepe University Journal of Sciences and Engineering. 2017;**17**:312-317. DOI: 10.5578/fmbd.53815

[24] Petrou C, Sarigiannis Y. Peptide synthesis. Peptide Applications in Biomedicine, Biotechnology and Bioengineering. 2018:1-21. DOI: 10.1016/B978-0-08-100736-5.00001-6

[25] Qvit N, Kornfeld OS. Development of a backbone cyclic peptide library as potential antiparasitic therapeutics using microwave irradiation. Journal of Visualized Experiments. 2016;(107):1-14. DOI: 10.3791/53589

[26] Fields GB. Introduction to peptide synthesis. Current Protocols in Protein Science. 2001;(18):1-9. DOI: 10.1002/0471140864.ps1801s26

[27] Skwarczynski M, Hussein WM, Liu T-Y, Toth I. Microwave-assisted synthesis of difficult sequence-containing peptide using the isopeptide method. Organic & Biomolecular Chemistry. 2013;**11**:2370-2376. DOI: 10.1039/c3ob00030c

[28] Tandrup Schmidt S, Foged C, Korsholm KS, Rades T, Christensen D.

Liposome-based adjuvants for subunit vaccines: Formulation strategies for subunit antigens and immunostimulators. Pharmaceutics. 2016;**8**(1):7. DOI: 10.3390/pharmaceutics8010007

[29] Palasek SA, Cox ZJ, Collins JM. Limiting racemization and aspartimide formation in microwave-enhanced fmoc solid phase peptide synthesis. Peptide Science. 2006;**13**:143-148. DOI: 10.1002/psc.804

[30] Moynihan JS et al. Enhanced immunogenicity of a hepatitis B virus peptide vaccine using oligosaccharide ester derivative microparticles. Vaccine. 2002;**20**(13-14):1870-1876. DOI: 10.1016/S0264-410X(01)00494-7

[31] Mustafaev MI, Mustafaeva Z. Novel polypeptide-comprising biopolymer system. Tecnology and Health Care. 2002;**10**:217-226

[32] Rafikov RZ et al. Pharmacokinetics of N-Vinylpyrrolidone copolymers with acrylic- acid. Khimiko-Farmatsevticheskii Zhurnal. 1986;**20**(3):272-274. DOI: 10.1007/BF00758558

[33] Ma W et al. PLGA nanoparticle-mediated delivery of tumor antigenic peptides elicits effective immune responses. International Journal of Nanomedicine. 2012;**7**:1475-1487. DOI: 10.2147/IJN.S29506

[34] Arasoglu T, Derman S, Mansuroglu B. Comparative evaluation of antibacterial activity of caffeic acid phenethyl ester and PLGA nanoparticle formulation by different methods. Nanotechnology. 2015;**27**:1-12. DOI: 10.3906/biy-1604-80

[35] Kızılbey K, Mansuroğlu B, Derman S, Mustafaeva Akdeste Z. An In vivo study: Adjuvant activity of poly-n-vinyl-2-pyrrolidone-co-acrylic acid on immune responses against melanoma synthetic peptide. Bioengineered. 2017;**9**(1):134-143. DOI: 10.1080/21655979.2017.1373529

[36] Mansuroglu B, Mustafaeva Z. Characterization of water-soluble conjugates of polyacrylic acid and antigenic peptide of FMDV by size exclusion chromatography with quadruple detection. Materials Science & Engineering, C: Materials for Biological Applications. 2012;**32**(2):112-118. DOI: 10.1016/j.msec.2011.10.004

[37] Pinilla-Ibarz J et al. Vaccination of patients with chronic myelogenous leukemia with bcr-abl oncogene breakpoint fusion peptides generates specific immune responses. Blood. 2000;**95**(5):1781-1787

[38] Kabanov VA. From synthetic polyelectrolytes to polymer-subunit vaccines. Pure and Applied Chemistry. 2004;**76**(9):1659-1677. DOI: 10.1351/pac200476091659

[39] Nair LS, Laurencin CT. Polymers as biomaterials for tissue engineering and controlled drug delivery. Tissue Engineering I: Scaffold Systems for Tissue Engineering. 2006;**102**:47-90. DOI: 10.1007/b137240

[40] Miladi K, Ibraheem D, Iqbal M, Sfar S, Fessi H, Elaissari A. Particles from preformed polymers as carriers for drug delivery. EXCLI Journal. 2014;**13**:28-57. DOI: 10.17877/DE290R-15560

[41] Sahle FF, Gerecke C, Kleuser B, Bodmeier R. Formulation and comparative in vitro evaluation of various dexamethasone-loaded pH-sensitive polymeric nanoparticles intended for dermal applications. International Journal of Pharmaceutics. 2017;**516**:21-31. DOI: 10.1016/j.ijpharm.2016.11.029

[42] Singh MN, Hemant KSY, Ram M, Shivakumar HG. Microencapsulation: A

promising technique for controlled drug delivery. Research in Pharmaceutical Sciences. 2010;**5**:65-77. License CC BY 2.0

[43] Acar T, Pelit Arayıcı P, Ucar B, Karahan M, Mustafaeva Z. Synthesis, characterization and lipophilicity study of Brucella abortus' immunogenic peptide sequence that can Be used in the future vaccination studies. International Journal of Peptide Research and Therapeutics. 2018:1-8. DOI: 10.1007/s10989-018-9739-0

[44] Karakus G, Polat AZ, Karahan M. Design, synthesis, structural characterization and cell cytotoxicity of a new derivative poly(maleic anhydride-co-vinyl acetate)/miltefosine polymer/drug conjugate. Bulgarian Chemical Communications. 2019;**51**(2):267-278. DOI: 10.34049/bcc.51.2.5053

[45] Çalman F, Pelit Arayıcı P, Büyükbayraktar HK, Karahan M, Mustafaeva Z, Katsarava R. Development of vaccine prototype against Zika virus disease of peptide-loaded PLGA nanoparticles and evaluation of cytotoxicity. International Journal of Peptide Research and Therapeutics. 2018:1-7. DOI: 10.1007/s10989-018-9753-2

[46] Mattson AM, Jenson CO, Dutcher RA. Triphenyltetrazolium as a dye for vital tissues. Science. 1947;**106**:294-295. DOI: 10.1126/science.106.2752.294-a

[47] Pagliacci MC, Spinozzi F, Migliorati G, Fumi G, Smacchia M, Grignani F, et al. Genistein inhibits tumour cell growth in vitro but enhances mitochondrial reduction of tetrazolium salts–A further pitfall in the use of the MTT assay for evaluating cell growth and survival. European Journal of Cancer. 1993;**29A**:1573-1577. DOI: 10.1016/0959-8049(93)90297-s

[48] Sereflioglu S, Yapici E, Tekarslan Sahin SH, Ozsoy Y, Ustundag CB. Targeted drug delivery and vaccinology approaches using virus-like particles for cancer. Istanbul Journal of Pharmacy. 2018;**47**(3):112-119. DOI: 10.5152/IstanbulJPharm.2017.0018

[49] Berridge MV, Herst PM, Tan AS. Tetrazolium dyes as tools in cell biology: New insights into their cellular reduction. Biotechnology Annual Review. 2005:127-152. DOI: 10.1016/S1387-2656(05)11004-7

[50] Wang L, Sun J, Horvat M, Koutalistras N, Johnston B, Ross Sheil AG. Evaluation of MTS, XTT, MTT and3HTdR incorporation for assessing hepatocyte density, viability and proliferation. Methods in Cell Science. 1996;**18**(3):249-255. DOI: 10.1007/BF00132890

[51] Saol Therapeutics Inc. Varizig: Highlights of prescribing information. The United States. Retrieved July 22, 2019. Available from: https://varizig.com/VARIZIG_PI.pdf

[52] Merck Sharp & Dohme Corp. Zostavax: Highlights of prescribing information. The United States. Retrieved July 22, 2019. Available from: https://www.merck.com/product/usa/pi_circulars/z/zostavax/zostavax_pi2.pdf

[53] Galetta KM, Gilden D. Zeroing in on zoster: A tale of many disorders produced by one virus. Journal of the Neurological Sciences. 2015;**358**(1-2):38-45. DOI: 10.1016/j.jns.2015.10.004

[54] Wang L, Zhu L, Zhu H. Efficacy of varicella (VZV) vaccination: An update for the clinician. Therapeutic Advances in Vaccines. 2016;**4**(1-2):20-31. DOI: 10.1177/2051013616655980

[55] Vartak A, Sucheck SJ. Recent advances in subunit vaccine carriers. Vaccine. 2016;**4**(2):12. DOI: 10.3390/vaccines4020012

[56] Henriksen-Lacey M, Korsholm KS, Andersen P, Perrie Y, Christensen D. Liposomal vaccine delivery systems. Expert Opinion on Drug Delivery. 2011;**8**(4):505-519. DOI: 10.1517/17425247.2011.558081

[57] Cataldi M, Cavaccini A. Eculizumab. In: xPharm: The Comprehensive Pharmacology Reference. Italy: Elsevier Inc.; 2010. pp. 1-26. DOI: 10.1016/B978-008055232-3.64509-X

[58] Benamu E, Montoya JG. Infections associated with the use of eculizumab. Current Opinion in Infectious Diseases. 2016;**29**(4):319-329. DOI: 10.1097/QCO.0000000000000279

[59] Rohovie MJ, Nagasawa M, Swartz JR. Virus-like particles: Next-generation nanoparticles for targeted therapeutic delivery. Bioengineering & Translational Medicine. 2017;**2**(1):43-57. DOI: 10.1002/btm2.10049

[60] Zhang L, Qiu W, Crooke S, Li Y, Abid A, Xu B, et al. Development of autologous C5 vaccine nanoparticles to reduce intravascular hemolysis in vivo. ACS Chemical Biology. 2017;**12**(2):539-547. DOI: 10.1021/acschembio.6b00994

[61] Pati R, Shevtsov M, Sonawane A. Nanoparticle vaccines against infectious diseases. Frontiers in Immunology. 2018;**9**:2224. DOI: 10.3389/fimmu.2018.02224

[62] Kilinc YB, Akdeste ZM, Koc RC, Bagirova M, Allahverdiyev A. Synthesis and characterization of antigenic influenza a M2e protein peptide-poly(acrylic) acid bioconjugate and determination of toxicityin vitro. Bioengineered. 2014;**5**(6):357-362. DOI: 10.4161/21655979.2014.969131

[63] Lai TKY, Su P, Zhang H, Liu F. Development of a peptide targeting dopamine transporter to improve ADHD-like deficits. Molecular Brain. 2018;**11**(1):1-14. DOI: 10.1186/s13041-018-0409-0

[64] Games D, Seubert P, Rockenstein E, Patrick C, Trejo M, Ubhi K, et al. Axonopathy in an α-Synuclein transgenic model of Lewy body disease is associated with extensive accumulation of C-terminal–truncated α-Synuclein. The American Journal of Pathology. 2013;**182**(3):940-953. DOI: 10.1016/J.AJPATH.2012.11.018

[65] Binukumar B, Shukla V, Amin ND, Grant P, Bhaskar M, Skuntz S, et al. Peptide TFP5/TP5 derived from Cdk5 activator P35 provides neuroprotection in the MPTP model of Parkinson's disease. Molecular Biology of the Cell. 2015;**26**(24):4478-4491. DOI: 10.1091/mbc.E15-06-0415

[66] Romero-Ramos M, von Euler Chelpin M, Sanchez-Guajardo V. Vaccination strategies for Parkinson disease. Human Vaccines & Immunotherapeutics. 2014;**10**(4):852-867. DOI: 10.4161/hv.28578

[67] Ghochikyan A, Petrushina I, Davtyan H, Hovakimyan A, Saing T, Davtyan A, et al. Immunogenicity of epitope vaccines targeting different B cell antigenic determinants of human α-synuclein: Feasibility study. Neuroscience Letters. 2014;**560**:86-91. DOI: 10.1016/j.neulet.2013.12.028

[68] Gozes I. Microtubules, schizophrenia and cognitive behavior: Preclinical development of davunetide (NAP) as a peptide-drug candidate. Peptides. 2011;**32**(2):428-431. DOI: 10.1016/j.peptides.2010.10.030

[69] Wang CY, Wang P-N, Chiu M-J, Finstad CL, Lin F, Lynn S, et al. UB-311, a novel UBITh ® amyloid β peptide vaccine for mild Alzheimer's

disease. Alzheimer's & Dementia: Translational Research & Clinical Interventions. 2017;**3**(2):262-272. DOI: 10.1016/j.trci.2017.03.005

[70] Wang Y, Li P, Truong-Dinh Tran T, Zhang J, Kong L. Manufacturing techniques and surface engineering of polymer based nanoparticles for targeted drug delivery to cancer. Nanomaterials. 2016;**6**(2):26. DOI: 10.3390/nano6020026

Section 3

Non Biological Based Nanomedicines

Chapter 4

Self-Emulsifying Drug Delivery Systems: Easy to Prepare Multifunctional Vectors for Efficient Oral Delivery

Khaled AboulFotouh, Ayat A. Allam and Mahmoud El-Badry

Abstract

Self-emulsifying drug delivery systems (SEDDS) have been mainly investigated to enhance the oral bioavailability of drugs belonging to class II of the Biopharmaceutics Classification System. However, in the past few years, they have shown promising outcomes in the oral delivery of various types of therapeutic agents. In this chapter, we discuss the recent progress in the application of SEDDS for oral delivery of protein therapeutics and genetic materials. The role of SEDDS in enhancing the oral bioavailability of P-glycoprotein and cytochrome P450 3A4 substrate drugs is also highlighted. Also, we discuss the most critical evaluation criteria of SEDDS. Additionally, we summarize various solidification techniques employed to transform liquid SEDDS to the more stable solid self-emulsifying drug delivery systems (s-SEDDS) that are associated with high patient compliance. This chapter provides a comprehensive approach to develop high utility SEDDS and their further transformation into s-SEDDS.

Keywords: solid self-emulsifying drug delivery systems, solidification techniques, oral delivery, P-glycoprotein (P-gp), cytochrome P450 3A4 (CYP3A4), multidrug resistance (MDR), protein therapeutics, plasmid DNA (pDNA)

1. Introduction

Lipid-based drug delivery systems (LBDDs) have been intensively investigated to overcome various obstacles encountered in oral drug delivery including poor aqueous solubility, limited permeability, low therapeutic window, first pass metabolism as well as inter- and intraindividual variability in drug response [1]. Lipid-based nanoparticles can achieve high loading capacity of hydrophilic and hydrophobic drugs [2]. The delivery features of these drug delivery systems could be tailored to achieve either immediate or sustained release properties depending on the appropriate selection of lipid composition. Most of lipids employed in the formulation are generally recognized as safe (GRAS), biocompatible and biodegradable [3]. LBDDs can enhance both transcellular and paracellular transport of drugs by transient disruption of lipid bilayer cells and alteration of tight junction by products of lipid digestion, respectively. Interestingly, they could permeate challenging physiological barriers such as blood brain barrier without surface modification due to their lipophilic nature [4]. Further, they are promising carriers for protection of therapeutic peptides against harsh GI environment [3]. Ease of preparation, cost effectiveness and possibility of large-scale production make LBDDs more attractive compared to polymeric nanoparticulate delivery systems [5].

2. Classification of lipid carriers

Lipid carriers can be classified into various categories depending mainly on their method of preparation as well as their physicochemical properties. They include liposomes, niosomes, solid lipid nanoparticles (SLNs), nanostructured lipid carriers (NLCs), micro and nanoemulsions, self-emulsifying drug delivery systems (SEDDS), and lipid-drug conjugates [2, 4].

Liposomes are uni- or multilamellar spherical vesicles which are composed of cholesterol and other natural or synthetic phospholipids enfolding an aqueous compartment [6]. They were first introduced by Bangham et al. in 1965 [7]. Thus, liposomes have been considered as biocompatible and biodegradable carriers that possess efficient delivering capability of hydrophilic and hydrophobic drugs. Advantages of liposome-based drug delivery systems include reduction of systemic and of target toxicities as well as targeting potential to achieve the desired outcome [8]. Thus, many liposome formulations have been approved for commercial use such Ambisome® (amphotericin B), Depocyt® (cytarabine), DepoDur® (morphine sulfate) and many others. However, their poor stability and rapid elimination by reticuloendothelial system limit the widespread applicability of liposomal formulations [4].

Niosomes are first described by Handjani-Vila et al. in 1979 [9]. They are nonionic surfactant-based vesicles in which the hydrophilic surfactant heads are oriented toward the exterior and the interior of the bilayer while, the hydrophobic tails are enclosed inside the bilayer. Therefore, like liposomes, niosomes have the ability to encapsulate hydrophilic or lipophilic molecules [10]. Niosomes also have cholesterol in their structure which enhances the rigidity of bilayer and reduces premature drug release [11]. Niosomes are superior carriers to liposomes in terms of production cost, chemical and physical stability, and loading capacity [12].

SLNs and NLCs are the most widely described solid-core lipid-based nanocarriers in the scientific literature [3]. SLNs were first described in 1991 to replace the liquid oil of O/W emulsions by a single solid lipid or mixture of solid lipids [2]. SLNs are composed of either solid lipid or mixture of lipids, that do not melt at room or physiological temperature, in an aqueous dispersion stabilized with the help of nonionic surfactants [3].SLNs offer the advantage of avoiding the use of organic solvents during preparation, effective delivery of both hydrophilic and lipophilic drugs, feasibility of surface functionalization with specific moieties to enhance their targeting potential, possibility of extended or controlled drug release, long shelf-life, biocompatibility, lower acute or chronic toxicity and effective large-scale production [2, 13, 14]. On the other hand, efficient drug delivery by SLNs is challenged by low drug loading capacity due to lipid crystalline nature, expulsion of loaded drug due to perfect crystalline lattice formation of the lipid and erratic gelation tendency that results in particle aggregation during storage [4, 14].

NLCs were developed to overcome the problem of drug expulsion during phase transition or crystallization of lipids comprising SLNs [15]. They also exist as a solid lipid matrix at temperature up to about 40°C. However, they are composed of solid lipid mixed with an oil which in turn reduces the lipid crystallization capacity and enhances the drug loading efficiency [16].

Nanoemulsions are kinetically stable heterogeneous systems composed of ultrafine oil droplets dispersed in aqueous media and stabilized by the aid of surfactants and cosurfactants. Nanoemulsions have gained increasing attention as promising drug delivery systems due to their multiple advantages including high surface area for drug absorption, biocompatibility, increasing drug solubility and improving mucosal permeability. Further, many FDA approved nanoemulsion-based products of water insoluble drugs are now available for clinical use including Restasis®, Estrasorb® and Flexogan®. Microemulsions also offer favorable characteristics such as thermodynamic stability, ease of production being formed spontaneously without the need for high energy input and high penetration due to the large surface area of internal phase [17].

Lipid-drug conjugates (LDCs) are lipid nanoparticle formulations which are characterized by the conjugation ability of the lipid matrix with the hydrophilic drug moieties, and thus provide novel pro-drugs to achieve many therapeutic outcomes in oral drug delivery [18]. Like other lipid-based nanocarrier systems, LDCs possess several advantages including biocompatibility, being solid at body and room temperature, high capacity for loading hydrophilic drugs, high permeation through GI tract, enhanced drug absorption through lymphatic uptake, improving stability and bioavailability loaded drugs, and feasibility of large scale production [19].

3. Self-emulsifying drug delivery systems

Self-emulsifying drug delivery systems (SEDDS) are lipid-based formulations that encompass isotropic mixtures of natural or synthetic oils, solid or liquid surfactants and co-surfactants [20]. When they are exposed to aqueous media (e.g., gastrointestinal fluids), they undergo self-emulsification to form O/W nanoemulsions or microemulsions with a mean droplet size between 20 and 200 nm [21]. Consequently, SEDDS are usually referred to as self-nanoemulsifying drug delivery systems (SNEDDS) or self-microemulsifying drug delivery systems (SMEDDS) depending on the nature of the resulting dispersions formed following their dilution [20].

SEDDS have been reported to enhance the oral bioavailability of poorly water-soluble drugs particularly those belonging to class II of the Biopharmaceutics Classification System by multiple underlying mechanisms [22]. Among these mechanisms, the enhanced drug solubilization was the most widely investigated. Lipidic components of SEDDS stimulate lipoprotein/chylomicron production thus promoting drug absorption [23]. The ultrafine droplet size range of the resulting emulsion provides a large surface area of interaction with gastrointestinal (GI) membranes [24]. Importantly, the bioactive effects of various ingredients employed in SEDDS formulation have significantly contributed to the enhanced oral bioavailability of the loaded drugs. These bioactive effects include tight junction opening and increasing membrane fluidity by the high surfactant content employed in SEDDS formulation [25]. Furthermore, stimulation of the intestinal lymphatic pathway as well as inhibition of intestinal drug efflux pumps such as P-glycoprotein (P-gp) and intestinal cytochrome P450 3A4 (CYP3A4) are considered promising strategies for enhancing the oral delivery of P-gp substrates and bypassing intestinal and hepatic first pass metabolism [26].

P-gp is an energy-dependent membrane bound protein and the most abundantly distributed ATP-binding cassette transmembrane transporter throughout the body [27]. P-gp prevents the accumulation of endogenous substances and xenobiotics in cells by transporting them back to the extracellular space [28]. Unfortunately, intestinal P-gp transporters hamper the intestinal uptake of substrate drugs thus, reducing their oral bioavailability. Additionally, overexpression of P-gp transporters is involved in the development of multidrug resistance (MDR) in numerous human tumor types [29]. Hence, many strategies have been developed to inhibit P-gp activity for enhancing the oral bioavailability of P-gp substrate drugs and reversing MDR in tumor cells. Among these strategies, nanocarriers have been widely investigated [26]. Nanocarriers have the advantage of protecting P-gp substrates against premature release and interaction with the biological environment [30]. They control drug tissue distribution and favorably accumulate in tumor tissue [31, 32]. Among various nanocarriers, SEDDS have been widely explored to enhance the oral bioavailability of P-gp substrate drugs and reverse MDR in tumor cells.

Interestingly, the unique combination of SEDDS properties allows the enhancement of oral bioavailability of both hydrophobic and hydrophilic drugs [33]. The oral delivery of protein therapeutics and genetic materials represents a real challenge due to their hydrophilic nature and their large molecular weight. In this chapter, we discuss the recent progress in the application of SEDDS for enhancing the oral bioavailability of P-gp substrates, reversing MDR in tumor cells and oral delivery of protein therapeutics and genetic materials. The aim of the current discussion is to call attention to the unique combination of SEDDS properties that makes them multifunctional delivery systems acting *via* various mechanisms to enhance the oral delivery of target therapeutic agents.

3.1 SEDDS overcome P-gp-mediated efflux and reverse MDR in tumor cells

Over the past 2 decades, SEDDS have been widely investigated to overcome P-gp-mediated efflux of substrate drugs to enhance their oral bioavailability. The potential of SEDDS to inhibit P-gp activity relies mainly on the excipients with established P-gp inhibition activity that are employed in the formulation [26]. Nonionic surfactants are the most widely employed excipients and are considered the mainstays of P-gp inhibition by SEDDS [29]. Cremophor EL, Cremophor RH40, vitamin E TPGS 1000, Labrasol, Transcutol P and Tween 80 are the most frequently employed. P-gp inhibition activity of a given surfactant depends on its HLB value and the structure of its hydrophobic domain [34]. There is no obvious correlation between surfactants' HLB values and P-gp inhibition activity. Structurally, the hydrophobic moieties of the surfactant should be linked to polyoxyethylene hydrophilic side chains to inhibit P-gp activity [34].

The binding affinity of nonionic surfactants to the hydrophobic portion of P-gp molecule is different from that of ionic surfactants [35]. Nonionic surfactants can change the secondary or tertiary structure of P-gp molecule resulting in the loss of its function [36]. Additionally, non-ionic surfactants were reported to modulate P-gp activity by inhibiting P-gp ATPase activity and either membrane fluidization or rigidization [37, 38]. At concentrations below the critical micelle concentration, nonionic surfactants are most effective in reducing P-gp activity; however, surfactant micelles showed some P-gp modulation activity [26].

SEDDS have superior formulation efficiency and *in vivo* performance compared to their individual components [39]. Various formulation aspects of SEDDS can potentiate the P-gp inhibition activity of their ingredients. The entrapment of P-gp substrate within the ultrafine emulsion droplets provides a protection against recognition by P-gp efflux pumps at GI epithelium [33]. In addition, SEDDS allow the co-administration of several excipients which are co-localized in close proximity to GI epithelium [22]. Further, pharmaceutical excipients with established P-gp inhibition activity (e.g., curcumin) or traditional P-gp inhibitors (e.g., elacridar) could be loaded into the SEDDS formulation to further augment their P-gp inhibition activity [40, 41].

On the other hand, the efflux of chemotherapeutic agents by P-gp transporters, which are overexpressed in tumor cells, represents a major obstacle in cancer chemotherapy [42]. SEDDS are extensively investigated to overcome MDR in tumor cells which is partly attributed to the overexpression of P-gp efflux transporters. SEDDS allow the combinational delivery of multiple chemotherapeutic agents acting *via* independent pathways in the same vector to produce a synergistic anticancer activity [42]. Interestingly, SEDDS could be employed for the co-delivery of various antioxidants for overcoming the oxidative stress in cancer cells [43].

3.2 SEDDS enhance the oral delivery of protein and peptide therapeutics

Protein therapeutics have a significant role in almost every field of medicine. However, the extensive application of protein therapeutics is challenged by their route of administration, being administered by parenteral route which is associated with reduced patient compliance [44]. Consequently, there is a great interest in the development of noninvasive strategies for delivery of protein therapeutics [45]. Oral delivery systems have been extensively investigated for the administration of protein drugs [46]. Unfortunately, the oral delivery of protein and/or peptide therapeutics is challenged by several barriers including the acidic environment

Protein	SEDDS composition	Bioavailability increase	Control	Animal species	Ref.
β-lactamase	Lauroglycol FCC (41.7%) Cremophor EL (33.3%) Transcutol HP (25%)	1.29-fold	β-lactamase solution	Sprague–Dawley rats	[54]
Insulin	Miglyol 840 (65%) Cremophor EL (25%) Co-solvent (DMSO and glycerol, 1:3) (10%)	3.33-fold	Insulin solution	Sprague–Dawley rats	[55]
Insulin	Ethyl oleate (35%) Cremophor El (32.5%) Alcohol (32.5%)	6.5-fold	Insulin solution	Male Wistar rats	[56]
Leuprorelin	Capmul MCM (30%) Cremophor EL (30%) Propylene glycol (10%) Captex 355 (30%)	17.2-fold	Leuprolide acetate solution	Sprague-Dawley rats	[57]
Pidotimod†	**Oil phase:** SoyPC (9.6%) Span 80 (21.1%) Oleic acid (36.1%) MCT (12%) 0.5% gelatin solution (3%) H_2O (12%) **Surfactant phase:** Tween 80 (6%)	2.56-fold	Pidotimod solution	Sprague–Dawley rats	[48]
Enoxaparin	Captex 8000 (30%) Capmul MCM (30%) Cremophor El (30%) Propylene glycol (10%)	2.25%▪	Enoxaparin IV solution	Sprague-Dawley rats	[58]
	Labrafil 1944 (35%) Capmul PG 8 (25%) Cremophor EL (30%) Propylene glycol (10%)	2.02%▪			

Abbreviations: DMSO, dimethyl sulfoxide; MCT, medium chain triglycerides.
▪*Absolute bioavailability.*
†*Self -double emulsifying drug delivery system.*

Table 1.
SEDDS-mediated enhancement in the oral bioavailability of various proteins. Reprinted with permission from Ref. [33] © Elsevier (2018).

in the stomach, degradation by GI enzymes, mucus barrier as well as low cellular penetration [47]. Several strategies have been developed to overcome these barriers [48–50]. As shown in **Table 1**, SEDDS have been extensively investigated as promising carriers for oral delivery of protein and peptide therapeutics. Various surfactants and oils that are employed in SEDDS formulation have a permeation enhancing effects; thus, they increase the cellular uptake of hydrophilic macromolecules such as protein therapeutics. The ultrafine droplet size provides a large surface area for rapid intestinal permeability. The anhydrous nature of SEDDS protects proteins against aqueous hydrolysis. Other bioactive effects of SEDDS such as tight junction opening, and enhanced lymphatic uptake also contribute to the enhanced oral bioavailability of loaded protein therapeutics [49]. However, loading of protein therapeutics into SEDDS is challenged by their hydrophilic nature. Thus, the lipid solubility of protein therapeutic should be increased before their incorporation into the SEDDS preconcentrate. This could be achieved by various techniques including, hydrophobic ion pairing [51], double emulsification [48], using hydrophilic solvents or co-solvents [52] and chemical modification of the peptide molecule [53].

Figure 1 summarizes various hypotheses for the enhanced oral delivery of protein and peptide therapeutics by SEDDS. Protein therapeutics incorporated within the ultrafine oil droplets are effectively protected against degradation by GI enzymes. Further, these cargoes are absorbed when the nanosized oil droplets are absorbed. Thus, the protection against enzymatic degradation is achieved *via* controlling the release rate of loaded protein therapeutic [59]. Burst release could result in rapid degradation of protein molecules within the GI lumen before reaching the absorption site [60]. Another suggested mechanism for the enhanced oral bioavailability by protein therapeutics incorporated in SEDDS is based on the bioactive effects of SEDDS ingredients. They include mucus penetration, enhanced paracellular transport *via* opening of tight junction, and enhanced cellular uptake by transcytosis-mediated transcellular transport [61, 62]. Finally, enhancing the lipid solubility of protein molecules *via* hydrophobic ion pairing could increase their intestinal uptake and bioavailability. However, this hypothesis is challenged by the rapid dissociation of hydrophobic ion paired complexes within the GI fluids.

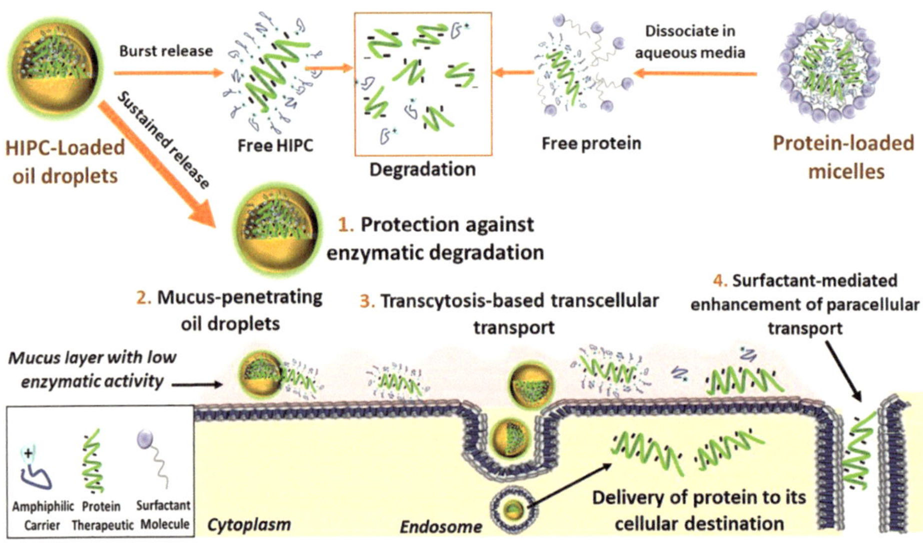

Figure 1.
A schematic representation of some underlying mechanisms for the enhanced oral bioavailability of protein therapeutics by SEDDS (HIPC, hydrophobic ion paired complex). Reprinted with permission from Ref. [33] © Elsevier (2018).

3.3 SEDDS as promising vectors for oral delivery of genetic materials

Oral gene therapy allows the sustained production of therapeutic proteins locally at the disease site as well as for systemic absorption [63]. Unfortunately, the oral delivery of plasmid DNA (pDNA) as well as other nucleic acid products is challenged by their safe and efficient delivery as well as cellular internalization and processing [64]. SEDDS have been investigated as promising non-viral vectors for oral delivery of genetic materials. The superior cellular permeation and stability of pDNA loaded into SEDDS could be mainly attributed to its entrapment within the ultrafine nanoemulsion oil droplets.

3.4 Characterization of SEDDS

3.4.1 Stability of SEDDS preconcentrates

SEDDS preconcentrates should have sufficient stability to avoid drug precipitation as well as creaming or phase separation of the resulting nano- or microemulsions. If some components of SEDDS preconcentrate undergo physical or chemical instability, the resulting emulsion may become unstable [20]. Thus, the stability of SEDDS preconcentrate should be evaluated by subjecting the nano- or microemulsion, resulting from aqueous dilution of the preconcentrate, to a centrifugation study at 5000 rpm for 30 min [65]. Then, SEDDS preconcentrates are subjected to heating–cooling cycle which includes six cycles of storage at 4 and 40°C for 48 h at each temperature followed by freeze–thaw cycle which involves three cycles of storage at −21 and 25°C for 48 h at each temperature [66].

3.4.2 Robustness to dilution

Robustness of the resulting emulsion to dilution guarantees the absence of drug precipitation when SEDDS preconcentrates are subjected to high dilution folds *in vivo* [21]. Thus, SEDDS preconcentrates should be exposed to different dilution folds (e.g., 50-, 100-, and 1000-folds) with different media (e.g., 0.1 N HCl and phosphate buffer, pH 6.8) to mimic *in vivo* conditions [20].

3.4.3 Assessment of self-emulsification efficiency

Self-emulsification efficiency is assessed by determining self-emulsification time and the efficiency of preconcentrate dispersibility when it is exposed to aqueous dilution. The SEDDS preconcentrate is added drop wise to aqueous media with different pH values and composition in a standard USP dissolution apparatus. Self-emulsification time is determined visually as the time required for the preconcentrate to form a homogenous dispersion [21]. The efficiency of preconcentrate dispersibility is also determined visually and is given in grades according to previously reported grading systems [20, 67, 68]. The selection of the appropriate grading system depends on the dilution fold to which the preconcentrate is exposed. This test ensures the ability of SEDDS preconcentrates to disperse quickly in order to form fine emulsions when they are exposed to aqueous media under mild agitation provided by the GI peristaltic movement.

3.4.4 Cloud point measurement

Cloud point could be measured after 100-fold dilution of the preconcentrate with distilled water which is then placed in a water bath with gradual increase

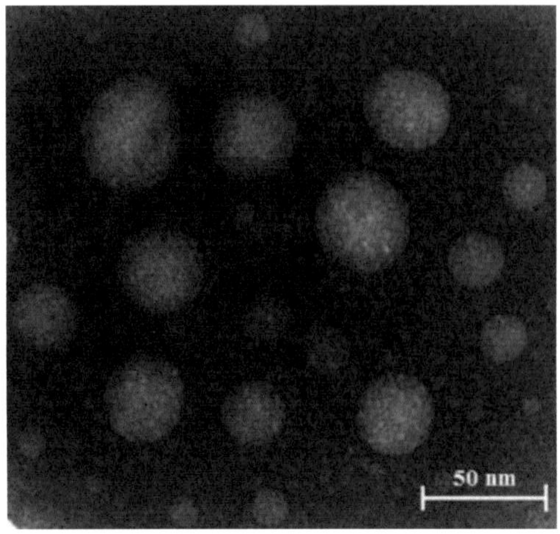

Figure 2.
Transmission electron microscope photograph shows spherical nanoemulsion droplets without any signs of aggregation. Reprinted with permission from Ref. [20]. © Elsevier (2017).

in temperature. Cloud point is determined as the temperature above which the emulsion clarity turns into cloudiness which is attributed to the dehydration of polyethylene oxide moiety of non-ionic surfactants [69]. Cloud point values should be sufficiently higher than 37°C (i.e., normal body temperature) to avoid phase separation in the GI tract [70].

3.4.5 Determination of zeta potential, mean droplet size and polydispersity index

Mean droplet size affects the *in vivo* performance of SEDDS. Small mean droplet size provides large interfacial area for drug absorption and ensures the kinetic stability of the resulting emulsion. Small value of polydispersity index suggests good uniformity of droplet size distribution. High zeta potential values confirm the electrical stability of emulsion droplets and absence of aggregation [66].

3.4.6 Droplet morphology

The morphology of emulsion droplets could be determined by transmission electron microscopy after appropriate dilution of SEDDS preconcentrate (about 1000-fold) using 2% solution of either phosphotungstic acid or uranyl acetate for negative staining. Droplets should possess a spherical shape without any signs of aggregation or drug precipitation as shown in **Figure 2**.

3.4.7 In vitro lipolysis

Drugs incorporated into lipid-based formulations are already present in a dissolved form. Thus, the assessment of the applicability of these formulations should be more properly based on the rate of drug precipitation over time. On the other hand, the drug solubilization capacity of lipid-based formulations is not a function of formulation characteristics alone. Rather, formulation dispersion and digestion result in the formation of colloidal species that account for the intestinal solubilization capacity [71]. Consequently, possible changes to solubilization capacity that could be attributed to digestion of formulation ingredients or interaction with biliary solubilizing agents should be assessed. *In vitro* lipolysis models simulate the GI environment and better predict the *in vivo* behavior of lipid-based formulations such as SEDDS. They also assess the extent of drug precipitation as a result of digestion of formulation ingredients and changes to solubilization capacity [72].

4. Solid self-emulsifying drug delivery systems

It was reported that ~70% of newly discovered drug molecules and ~ 40% of marketed drugs for oral administration are classified as practically insoluble in water. Therefore, various strategies have been explored to enhance the aqueous solubility and thus oral bioavailability of these drugs. SEDDS have been investigated as an efficient strategy that allows drug administration in pre-solubilized form ready for absorption. Consequently, drugs loaded into SEDDS preconcentrates avoid the dissolution step that frequently limits their absorption. However, the widespread application of liquid SEDDS is challenged by low stability during handling or storage [73] and irreversible drug and/or excipient precipitation [74]. Thus, the majority of marketed liquid SEDDS are filled into soft gelatin (e.g., SandimunneNeoral®, Norvir®, Fortovase®, and Convulex®) or hard gelatin capsules (e.g., Gengraf® and Lipirex®) to be administered as a unit dosage form [75]. However, this approach still possesses the possibility of drug precipitation upon exposure to aqueous media. Additionally, capsule technologies have some limitations such as high production cost and the risk of interaction between the active pharmaceutical ingredient and excipients with the capsule shell. Also, the possibility of drug leakage out of the capsule shell and capsule aging represent further obstacles [76]. Further, the storage temperature is an important consideration since the drug and/or excipients could undergo precipitation at lower temperatures [75]. The tendency of volatile excipients to evaporate into the capsule shell results in drug precipitation and a consequent alteration of drug release [77].

Thus, to address these limitations solid self-emulsifying drug delivery systems (s-SEDDS) were developed by converting the conventional liquid SEDDS into powders which are subsequently filled into capsules or formulated as solid dosage forms such as self-emulsifying tablets, granules, pellets, beads, microspheres, nanoparticles, suppositories and implants [74, 78]. Various solidification techniques for converting liquid SEDDS into s-SEDDS are discussed below.

4.1 Solidification techniques for converting liquid or semisolid SEDDS to s-SEDDS

4.1.1 Adsorption to solid carriers

Adsorption to highly porous and/or high specific area solid carriers is the most intensively explored approach to obtain s-SEDDS [75]. This technique could be effectively used to produce s-SEDDS by simple mixing of solid carriers with the liquid formulation in a blender [74]. The most frequently employed carriers for adsorption of liquid SEDDS formulations are: (i) silicon dioxide such as Aerosil® (fumed silica) and Sylysia® (micronized amorphous silica); (ii) Neusilin® (magnesium aluminometasilicate) which is available in different surface properties and particle size; (iii) Fujicalin® (porous dibasic calcium phosphate anhydrous) and (iv) calcium silicate [75].

Advantages of this solidification technique include:(i) good content uniformity of the produced powders [79]; (ii) high drug loading efficiency (up to 80% *w/w* without affecting flow properties) [80]; (iii) absence of organic solvents [81]; (iv) cost effectiveness because small number of excipients and basic equipment are required for the final formulation and (v) production of free-flowing powders that can be filled into capsule or compressed into other solid dosage form [82].

During the formulation of s-SEDDS by adsorption technique, careful consideration should be given to the possible interactions between the solid carrier and the drug or other excipients in liquid SEDDS which could result in delayed or incomplete release of loaded drug [83]. Additionally, the particle size, specific surface area, tortuosity of pores as well as type and liquid SEDDS: carrier ratio should be considered [75].

4.1.2 Spray drying

Spray drying is also a promising technique for transforming liquid SEDDS to s-SEDDS using different carriers (i.e., hydrophobic or hydrophilic carriers) which preserve the self-emulsifying properties of the formulation. It is a simple and economical technique which involves mixing of lipids, surfactants, drug and solid carriers followed by solubilization and spray drying. The solubilized mixture is atomized into a spray of fine droplets that are introduced into a drying chamber where the volatile phase evaporates forming dry particles under controlled conditions of temperature and airflow [74]. The type of carrier can affect the rate of release and thus the oral bioavailability of loaded drug by affecting the droplet size of the nano or microemulsion formed after reconstitution [84]. Also, careful consideration should be given to the atomizer, the airflow pattern, the temperature and the design of the drying chamber which should be selected according to the powder specifications. Low yield is a disadvantage of solidification by spray drying technique which could be attributed to the removal of non-encapsulated drug with the exhausted air [85].

4.1.3 Extrusion/spheronization

Extrusion/spheronization is the most explored technique for the production of uniformly sized self-emulsifying pellets [75]. Extrusion is a procedure of converting a raw material with plastic properties into a spaghetti-shaped agglomerate having uniform density. Extrusion is followed by spheronization where the extrudate is broken into spherical pellets (spheroids) of uniform size [86]. The produced pellets have good flowability and low friability. Before pellet production, the wet mass is composed of liquid SEDDS, lactose, microcrystalline cellulose (MCC) and water. A disintegrating agent could be added to enhance drug release [87]. MCC acts as adsorbent for the liquid SEDDS to ease pellet formation and avoid problems such as poor flow properties, pellet agglomeration and low hardness. Larger amount of liquid SEDDS can be loaded into the pellets when a greater quantity of MCC on the account of lower amount of lactose is employed in the formulation. The ratio of lactose: MCC and liquid SEDDS: water affects the pellets' disintegration time and surface roughness as well as the extrusion force [88].

4.1.4 Microencapsulation

Co-extrusion technique is a promising strategy for microencapsulation of liquid SEDDS into polymeric matrices. This technique employs a vibrating nozzle device equipped with a concentric nozzle. The formed microcapsules are then hardened by ionotropic gelation. Ionotropic gelation is based on the gel formation ability of polysaccharides (e.g., pectin, alginate, carrageenan, and gellan) in the presence of multivalent ions (e.g., Ca^{+2}) [89]. Alginate and pectin are the most intensively investigated natural ionic polysaccharides for formation of microcapsule shell. However, Ca-alginate microcapsules clog the nozzle during the microencapsulation process. On the other hand, pectin microcapsules lack sufficient hardness. Thus, microcapsules composed of an alginate-pectin matrix could be more acceptable than those composed solely of one polymer. Various hydrophilic filling agents (e.g., lactose) could be added to the shell formation phase in order to prevent core leakage and microcapsule collapse during the drying process. Advantages of microcapsules include predictable GI transit time and large surface area that allow faster drug dissolution. Additionally, they are composed of biocompatible, non-toxic and biodegradable natural polymers [90].

4.1.5 Wet granulation

Different carriers (e.g., Aerosil® 200) were employed to prepare the self-emulsifying granules where the liquid SEDDS acts as a binder. However, granulation with SEDDS produces a broader size distribution and difficult to control aggregation compared with granulation procedure where water is employed as granulating agent [91].

4.1.6 Melt granulation

In this process, powder agglomeration is attained by the addition of binding agent which melts at relatively low temperature such as Gelucire®, lecithin, partial glycerides or polysorbates [92]. While the liquid SEDDS is adsorbed to neutral carriers such as silica and magnesium aluminometasilicate [93]. Melt granulation is advantageous compared to wet granulation since it is a 'one-step' process in which the addition of granulating liquid and the following drying phase are absent [74].

4.2 Characterization of s-SEDDS

SEDDS are combinations of SEDDS and solid dosage forms. Therefore, the characterization of s-SEDDS is the sum of the corresponding evaluation criteria of both SEDDS and solid dosage forms.

4.2.1 Solid state characterization

4.2.1.1 Differential scanning calorimetry (DSC)

DSC is mainly employed to ensure drug incorporation into the s-SEDDS as well as the absence of drug-solid carrier interaction. It is also used to investigate the physical state (i.e., crystalline or amorphous) of the incorporated drug in the final formulation [94]. Transition from the crystalline to amorphous state is common in SEDDS formulations which lowers the drug melting point and improves its solubility and dissolution rate [95].

4.2.1.2 X-ray diffractometry (XRD)

XRD is employed to investigate the physical state of the incorporated drug because it affects both *in vitro* and *in vivo* performance.

4.2.1.3 Scanning electron microscopy (SEM)

SEM is employed to elucidate the structural and morphological features of s-SEDDS and the raw materials as well as to confirm the physical state of loaded drug [96].

4.2.1.4 Fourier-transform infrared spectroscopy (FT-IR)

FT-IR is usually employed to investigate any potential interaction between the incorporated drug and the solid carrier or other formulation excipients [76].

4.2.2 Determination of micromeritic properties

The flow properties of powders are crucial aspect of large-scale production of solid dosage forms because it affects feeding consistency, reproducibility of die filling and dose uniformity. Powder flowability is affected by various physical, mechanical and environmental factors. Thus, various parameters such as angle of repose, bulk density, Carr's index and Hausner's ratio should be assessed to determine s-SEDDS flowability to overcome the subjective nature of individual tests. The angle of repose is a measure of internal cohesiveness of particles. Powders having angles of repose <30° are considered as free flowing powders; while, powders with angles of repose >40° are regarded to have extremely poor flowability. On the other hand, powders with angles of repose up to 35° are regarded passable; while, those between 35 and 40° indicate poor powder flow which requires the addition of a glidant [97]. Powders having Carr's index up to 21% are considered to have acceptable flow. Hausner's ratios <1.25 are usually corresponded to free-flowing powders with minimum interparticle frictions. On the other hand, Hausner's ratios between 1.25 and 1.5 indicate moderate flow which could be acceptable [98].

4.2.3 Droplet size of reconstituted s-SEDDS

The droplet size of reconstituted s-SEDDS should be similar to that of liquid SEDDS to ensure that the self-emulsification performance of liquid SEDDS is preserved.

5. Conclusion

SEDDS are promising nanocarriers for overcoming various obstacles encountered in the oral delivery of drugs and bioactive agents. The inhibition of P-gp activity by SEDDS relies mainly on the employment of ingredients (i.e., oils and surfactants) with established P-gp inhibition activity in their formulation. Thus, selection of excipients with established P-gp inhibition activity is the first step in the formulation of SEDDS for overcoming P-gp-mediated efflux of substrate drugs and reversing MDR in tumor cells. The effective concentration range for inhibiting P-gp activity should be considered while selecting the formulation ratios. P-gp inhibition activity of SEDDS can be further enhanced by loading other pharmaceutical excipient with established P-gp inhibition activity or traditional P-gp inhibitor. SEDDS are also considered promising systems for the oral delivery of protein therapeutics and genetic materials; however, this role is still in its infancy. Entrapment of these macromolecules within the nanosized emulsion droplets guarantees effective delivery. The bioactive effects of SEDDS ingredients could further enhance the oral bioavailability of protein therapeutics. Liquid SEDDS could be transformed into s-SEDDS to further enhance the formulation stability, allow cost effective large-scale production as well as to enhance the patient compliance.

Conflict of interest

None.

Author details

Khaled AboulFotouh, Ayat A. Allam and Mahmoud El-Badry*
Department of Pharmaceutics, Faculty of Pharmacy, Assiut University, Assiut, Egypt

*Address all correspondence to: elbadry@aun.edu.eg

References

[1] Swarnakar NK, Venkatesan N, Betageri G. Critical In vitro characterization methods of lipid-based formulations for oral delivery: A comprehensive review. AAPS PharmSciTech. 2018;**20**(1):16. DOI: 10.1208/s12249-018-1239-1

[2] Talegaonkar S, Bhattacharyya A. Potential of lipid nanoparticles (SLNs and NLCs) in enhancing oral bioavailability of drugs with poor intestinal permeability. AAPS PharmSciTech. 2019;**20**(3):121. DOI: 10.1208/s12249-019-1337-8

[3] Dumont C et al. Lipid-based nanosuspensions for oral delivery of peptides: A critical review. International Journal of Pharmaceutics. 2018;**541**(1):117-135. DOI: 10.1016/j.ijpharm.2018.02.038

[4] Khosa A, Reddi S, Saha RN. Nanostructured lipid carriers for site-specific drug delivery. Biomedicine & Pharmacotherapy. 2018;**103**:598-613. DOI: 10.1016/j.biopha.2018.04.055

[5] Tapeinos C, Battaglini M, Ciofani G. Advances in the design of solid lipid nanoparticles and nanostructured lipid carriers for targeting brain diseases. Journal of Controlled Release. 2017;**264**:306-332. DOI: 10.1016/j.jconrel.2017.08.033

[6] Ahmed KS et al. Liposome: Composition, characterisation, preparation, and recent innovation in clinical applications. Journal of Drug Targeting. 2019;**27**(7):742-761. DOI: 10.1080/1061186X.2018.1527337

[7] Bangham AD, Standish MM, Watkins JC. Diffusion of univalent ions across the lamellae of swollen phospholipids. Journal of Molecular Biology. 1965;**13**(1):238-IN27. DOI: 10.1016/S0022-2836(65)80093-6

[8] Zahednezhad F et al. Liposome and immune system interplay: Challenges and potentials. Journal of Controlled Release. 2019;**305**:194-209. DOI: 10.1016/j.jconrel.2019.05.030

[9] Handjani-Vila R et al. Dispersions of lamellar phases of non-ionic lipids in cosmetic products. International Journal of Cosmetic Science. 1979;**1**(5):303-314

[10] Verma S, Utreja P. Vesicular nanocarrier based treatment of skin fungal infections: Potential and emerging trends in nanoscale pharmacotherapy. Asian Journal of Pharmaceutical Sciences. 2019;**14**(2):117-129. DOI: 10.1016/j.ajps.2018.05.007

[11] Abdelkader H, Alani AWG, Alany RG. Recent advances in non-ionic surfactant vesicles (niosomes): Self-assembly, fabrication, characterization, drug delivery applications and limitations. Drug Delivery. 2014;**21**(2):87-100. DOI: 10.3109/10717544.2013.838077

[12] Azeem A, Anwer MK, Talegaonkar S. Niosomes in sustained and targeted drug delivery: Some recent advances. Journal of Drug Targeting. 2009;**17**(9):671-689. DOI: 10.3109/10611860903079454

[13] Ganesan P et al. Recent developments in solid lipid nanoparticle and surface-modified solid lipid nanoparticle delivery systems for oral delivery of phyto-bioactive compounds in various chronic diseases. International Journal of Nanomedicine. 2018;**13**:1569-1583. DOI: 10.2147/IJN.S155593

[14] Nabi B et al. Insights on oral drug delivery of lipid nanocarriers: A win-win solution for augmenting bioavailability of antiretroviral drugs.

AAPS PharmSciTech. 2019;**20**(2):60. DOI: 10.1208/s12249-018-1284-9

[15] Almeida AJ, Souto E. Solid lipid nanoparticles as a drug delivery system for peptides and proteins. Advanced Drug Delivery Reviews. 2007;**59**(6):478-490. DOI: 10.1016/j.addr.2007.04.007

[16] Weber S, Zimmer A, Pardeike J. Solid lipid nanoparticles (SLN) and nanostructured lipid carriers (NLC) for pulmonary application: A review of the state of the art. European Journal of Pharmaceutics and Biopharmaceutics. 2014;**86**(1):7-22. DOI: 10.1016/j.ejpb.2013.08.013

[17] Siqueira Leite CB et al. Microemulsions as platforms for transdermal delivery of hydrophilic drugs-a review. Current Nanoscience. 2018;**14**(3):170-178

[18] Attama AA, Momoh MA, Builders PF. Lipid nanoparticulate drug delivery systems: A revolution in dosage form design and development. In: Recent Advances in Novel Drug Carrier Systems. London, UK: IntechOpen; 2012

[19] Banerjee S, Kundu A. Lipid-drug conjugates: A potential nanocarrier system for oral drug delivery applications. DARU Journal of Pharmaceutical Sciences. 2018;**26**(1): 65-75. DOI: 10.1007/s40199-018-0209-1

[20] AboulFotouh K et al. Development and in vitro/in vivo performance of self-nanoemulsifying drug delivery systems loaded with candesartan cilexetil. European Journal of Pharmaceutical Sciences. 2017;**109**:503-513. DOI: 10.1016/j.ejps.2017.09.001

[21] Balakumar K et al. Self nanoemulsifying drug delivery system (SNEDDS) of Rosuvastatin calcium: Design, formulation, bioavailability and pharmacokinetic evaluation. Colloids and Surfaces B: Biointerfaces. 2013;**112**:337-343. DOI: 10.1016/j.colsurfb.2013.08.025

[22] Elgart A et al. Improved oral bioavailability of BCS class 2 compounds by self nano-emulsifying drug delivery systems (SNEDDS): The underlying mechanisms for amiodarone and talinolol. Pharmaceutical Research. 2013;**30**(12):3029-3044. DOI: 10.1007/s11095-013-1063-y

[23] Kale AA, Patravale VB. Design and evaluation of self-emulsifying drug delivery systems (SEDDS) of nimodipine. AAPS PharmSciTech. 2008;**9**(1):191-196. DOI: 10.1208/s12249-008-9037-9

[24] Dash RN et al. Solid supersaturatable self-nanoemulsifying drug delivery systems for improved dissolution, absorption and pharmacodynamic effects of glipizide. Journal of Drug Delivery Science and Technology. 2015;**28**:28-36. DOI: 10.1016/j.jddst.2015.05.004

[25] Sun M et al. The influence of co-solvents on the stability and bioavailability of rapamycin formulated in self-microemulsifying drug delivery systems. Drug Development and Industrial Pharmacy. 2011;**37**(8):986-994. DOI: 10.3109/03639045.2011.553618

[26] AboulFotouh K et al. Self-emulsifying drug–delivery systems modulate P-glycoprotein activity: Role of excipients and formulation aspects. Nanomedicine. 2018;**13**(14):1813-1834. DOI: 10.2217/nnm-2017-0354

[27] Karan M, Rajashree CM, Arti RT. Challenges in Oral delivery: Role of P-gp efflux pump. Current Drug Therapy. 2014;**9**(1):47-55. DOI: 10.2174/1574885509666140805003456

[28] Yu J et al. Advances in plant-based inhibitors of P-glycoprotein. Journal of Enzyme Inhibition and Medicinal Chemistry. 2016;**31**(6):867-881. DOI: 10.3109/14756366.2016.1149476

[29] Sosnik A. Reversal of multidrug resistance by the inhibition of ATP-binding cassette pumps employing "generally recognized as safe" (GRAS) nanopharmaceuticals: A review. Advanced Drug Delivery Reviews. 2013;**65**(13):1828-1851. DOI: 10.1016/j.addr.2013.09.002

[30] Kaur V et al. Therapeutic potential of nanocarrier for overcoming to P-glycoprotein. Journal of Drug Targeting. 2014;**22**(10):859-870. DOI: 10.3109/1061186X.2014.947295

[31] Kapse-Mistry S et al. Nanodrug delivery in reversing multidrug resistance in cancer cells. Frontiers in Pharmacology. 2014;**5**(159):1-22. DOI: 10.3389/fphar.2014.00159

[32] Kim C-K, Lim S-J. Recent progress in drug delivery systems for anticancer agents. Archives of Pharmacal Research. 2002;**25**(3):229. DOI: 10.1007/bf02976620

[33] AboulFotouh K et al. Role of self-emulsifying drug delivery systems in optimizing the oral delivery of hydrophilic macromolecules and reducing interindividual variability. Colloids and Surfaces B: Biointerfaces. 2018;**167**:82-92. DOI: 10.1016/j.colsurfb.2018.03.034

[34] Dewanjee S et al. Natural products as alternative choices for P-glycoprotein (P-gp) inhibition. Molecules. 2017;**22**(6):871

[35] Shono Y et al. Modulation of intestinal P-glycoprotein function by Cremophor EL and other surfactants by an in vitro diffusion chamber method using the isolated rat intestinal membranes. Journal of Pharmaceutical Sciences. 2004;**93**(4):877-885. DOI: 10.1002/jps.20017

[36] Akhtar N et al. The emerging role of P-glycoprotein inhibitors in drug delivery: A patent review. Expert Opinion on Therapeutic Patents. 2011;**21**(4):561-576. DOI: 10.1517/13543776.2011.561784

[37] Rege BD, Kao JPY, Polli JE. Effects of nonionic surfactants on membrane transporters in Caco-2 cell monolayers. European Journal of Pharmaceutical Sciences. 2002;**16**(4):237-246. DOI: 10.1016/S0928-0987(02)00055-6

[38] Zhao W et al. Effects of 2 polyoxyethylene alkyl ethers on the function of intestinal P-glycoprotein and their inhibitory mechanisms. Journal of Pharmaceutical Sciences. 2016;**105**(12):3668-3679. DOI: 10.1016/j.xphs.2016.09.002

[39] Hintzen F et al. In vitro and ex vivo evaluation of an intestinal permeation enhancing self-microemulsifying drug delivery system (SMEDDS). Journal of Drug Delivery Science and Technology. 2013;**23**(3):261-267. DOI: 10.1016/S1773-2247(13)50039-6

[40] Sandhu PS et al. Novel dietary lipid-based self-nanoemulsifying drug delivery systems of paclitaxel with P-gp inhibitor: Implications on cytotoxicity and biopharmaceutical performance. Expert Opinion on Drug Delivery. 2015;**12**(11):1809-1822. DOI: 10.1517/17425247.2015.1060219

[41] Chaurasiya A et al. Dual approach utilizing self microemulsifying technique and novel P-gp inhibitor for effective delivery of taxanes. Journal of Microencapsulation. 2012;**29**(6):583-595. DOI: 10.3109/02652048.2012.668959

[42] Saneja A et al. 14—Recent advances in self-emulsifying drug-delivery systems for oral delivery of cancer chemotherapeutics A2—Holban, Alina Maria. In: Grumezescu AM, editor. Nanoarchitectonics for Smart Delivery and Drug Targeting. Norwich, New York, USA: William Andrew Publishing; 2016. pp. 379-404

[43] Jain S et al. α-Tocopherol as functional excipient for resveratrol and coenzyme Q10-loaded SNEDDS for improved bioavailability and prophylaxis of breast cancer. Journal of Drug Targeting. 2017;**25**(6):554-565. DOI: 10.1080/1061186X.2017.1298603

[44] Leader B, Baca Q J, Golan DE. Protein therapeutics: A summary and pharmacological classification. Nature Reviews. Drug Discovery. 2008;**7**(1):21-39

[45] Ijaz M et al. Development of oral self nano-emulsifying delivery system(s) of lanreotide with improved stability against presystemic thiol-disulfide exchange reactions. Expert Opinion on Drug Delivery. 2016;**13**(7):923-929. DOI: 10.1517/17425247.2016.1167034

[46] Karamanidou T et al. Effective incorporation of insulin in mucus permeating self-nanoemulsifying drug delivery systems. European Journal of Pharmaceutics and Biopharmaceutics. 2015;**97**:223-229. DOI: 10.1016/j.ejpb.2015.04.013

[47] Leonaviciute G, Bernkop-Schnürch A. Self-emulsifying drug delivery systems in oral (poly)peptide drug delivery. Expert Opinion on Drug Delivery. 2015;**12**(11):1703-1716. DOI: 10.1517/17425247.2015.1068287

[48] Qi X et al. Self-double-emulsifying drug delivery system (SDEDDS): A new way for oral delivery of drugs with high solubility and low permeability. International Journal of Pharmaceutics. 2011;**409**(1):245-251. DOI: 10.1016/j.ijpharm.2011.02.047

[49] Li P, Nielsen HM, Müllertz A. Impact of lipid-based drug delivery systems on the transport and uptake of insulin across Caco-2 cell monolayers. Journal of Pharmaceutical Sciences. 2016;**105**(9):2743-2751. DOI: 10.1016/j.xphs.2016.01.006

[50] Li P et al. Self-nanoemulsifying drug delivery systems for oral insulin delivery: In vitro and in vivo evaluations of enteric coating and drug loading. International Journal of Pharmaceutics. 2014;**477**(1):390-398. DOI: 10.1016/j.ijpharm.2014.10.039

[51] Zhang J et al. Preparation, characterization, and in vivo evaluation of a self-nanoemulsifying drug delivery system (SNEDDS) loaded with morin-phospholipid complex. International Journal of Nanomedicine. 2011;**6**:3405-3414. DOI: 10.2147/IJN.S25824

[52] Winarti L. Formulation of self-nanoemulsifying drug delivery system of bovine serum albumin using HLB (hydrophilic-lypophilic balance) approach. Indonesian Journal of Pharmacy. 2016;**27**(3):117-127

[53] Zupančič O et al. Development and in vitro characterization of self-emulsifying drug delivery system (SEDDS) for oral opioid peptide delivery. Drug Development and Industrial Pharmacy. 2017;**43**(10):1-9. DOI: 10.1080/03639045.2017.1338722

[54] Rao SVR, Yajurvedi K, Shao J. Self-nanoemulsifying drug delivery system (SNEDDS) for oral delivery of protein drugs: III. In vivo oral absorption study. International Journal of Pharmaceutics. 2008;**362**(1):16-19

[55] Sakloetsakun D et al. Combining two technologies: Multifunctional polymers and self-nanoemulsifying drug delivery system (SNEDDS) for oral insulin administration. International Journal of Biological Macromolecules. 2013;**61**:363-372. DOI: 10.1016/j.ijbiomac.2013.08.002

[56] Zhang Q et al. The In vitro and In vivo study on self-nanoemulsifying drug delivery system (SNEDDS) based on insulin-phospholipid complex. Journal of Biomedical Nanotechnology.

2012;**8**(1):90-97. DOI: 10.1166/jbn.2012.1371

[57] Hintzen F et al. In vivo evaluation of an oral self-microemulsifying drug delivery system (SMEDDS) for leuprorelin. International Journal of Pharmaceutics. 2014;**472**(1):20-26. DOI: 10.1016/j.ijpharm.2014.05.047

[58] Zupančič O et al. Development, in vitro and in vivo evaluation of a self-emulsifying drug delivery system (SEDDS) for oral enoxaparin administration. European Journal of Pharmaceutics and Biopharmaceutics. 2016;**109**:113-121. DOI: 10.1016/j.ejpb.2016.09.013

[59] Lupo N et al. Inhibitory effect of emulsifiers in SEDDS on protease activity: Just an illusion? International Journal of Pharmaceutics. 2017;**526**(1):23-30. DOI: 10.1016/j.ijpharm.2017.04.058

[60] Soltani Y, Goodarzi N, Mahjub R. Preparation and characterization of self nano-emulsifying drug delivery system (SNEDDS) for oral delivery of heparin using hydrophobic complexation by cationic polymer of β-cyclodextrin. Drug Development and Industrial Pharmacy. 2017;**43**(11):1-9. DOI: 10.1080/03639045.2017.1353522

[61] Ma E-L et al. In vitro and in vivo evaluation of a novel oral insulin formulation. Acta Pharmacologica Sinica. 2006;**27**(10):1382-1388. DOI: 10.1111/j.1745-7254.2006.00424.x

[62] Lindmark T, Nikkilä T, Artursson P. Mechanisms of absorption enhancement by medium chain fatty acids in intestinal epithelial Caco-2 cell monolayers. Journal of Pharmacology and Experimental Therapeutics. 1995;**275**(2):958-964

[63] Bhavsar MD, Amiji MM. Polymeric nano- and microparticle technologies for oral gene delivery. Expert Opinion on Drug Delivery. 2007;**4**(3):197-213. DOI: 10.1517/17425247.4.3.197

[64] Guliyeva Ü et al. Chitosan microparticles containing plasmid DNA as potential oral gene delivery system. European Journal of Pharmaceutics and Biopharmaceutics. 2006;**62**(1):17-25. DOI: 10.1016/j.ejpb.2005.08.006

[65] Negi LM, Tariq M, Talegaonkar S. Nano scale self-emulsifying oil based carrier system for improved oral bioavailability of camptothecin derivative by P-glycoprotein modulation. Colloids and Surfaces B: Biointerfaces. 2013;**111**:346-353. DOI: 10.1016/j.colsurfb.2013.06.001

[66] Parmar N et al. Study of cosurfactant effect on nanoemulsifying area and development of lercanidipine loaded (SNEDDS) self nanoemulsifying drug delivery system. Colloids and Surfaces B: Biointerfaces. 2011;**86**(2):327-338. DOI: 10.1016/j.colsurfb.2011.04.016

[67] Balakumar K et al. Self emulsifying drug delivery system: Optimization and its prototype for various compositions of oils, surfactants and co-surfactants. Journal of Pharmacy Research. 2013;**6**(5):510-514. DOI: 10.1016/j.jopr.2013.04.031

[68] Khoo S-M et al. Formulation design and bioavailability assessment of lipidic self-emulsifying formulations of halofantrine. International Journal of Pharmaceutics. 1998;**167**(1):155-164. DOI: 10.1016/S0378-5173(98)00054-4

[69] Elnaggar YSR, El-Massik MA, Abdallah OY. Self-nanoemulsifying drug delivery systems of tamoxifen citrate: Design and optimization. International Journal of Pharmaceutics. 2009;**380**(1):133-141. DOI: 10.1016/j.ijpharm.2009.07.015

[70] Bandyopadhyay S, Katare OP, Singh B. Optimized self nano-emulsifying

systems of ezetimibe with enhanced bioavailability potential using long chain and medium chain triglycerides. Colloids and Surfaces B: Biointerfaces. 2012;**100**:50-61. DOI: 10.1016/j.colsurfb.2012.05.019

[71] Porter CJH et al. Enhancing intestinal drug solubilisation using lipid-based delivery systems. Advanced Drug Delivery Reviews. 2008;**60**(6):673-691. DOI: 10.1016/j.addr.2007.10.014

[72] Larsen AT et al. Oral bioavailability of cinnarizine in dogs: Relation to SNEDDS droplet size, drug solubility and in vitro precipitation. European Journal of Pharmaceutical Sciences. 2013;**48**(1):339-350. DOI: 10.1016/j.ejps.2012.11.004

[73] Mustapha O et al. Development of novel cilostazol-loaded solid SNEDDS using a SPG membrane emulsification technique: Physicochemical characterization and in vivo evaluation. Colloids and Surfaces B: Biointerfaces. 2017;**150**:216-222. DOI: 10.1016/j.colsurfb.2016.11.039

[74] Tang B et al. Development of solid self-emulsifying drug delivery systems: Preparation techniques and dosage forms. Drug Discovery Today. 2008;**13**(13):606-612. DOI: 10.1016/j.drudis.2008.04.006

[75] Mandić J et al. Overview of solidification techniques for self-emulsifying drug delivery systems from industrial perspective. International Journal of Pharmaceutics. 2017;**533**(2):335-345. DOI: 10.1016/j.ijpharm.2017.05.036

[76] Sanka K, Suda D, Bakshi V. Optimization of solid-self nanoemulsifying drug delivery system for solubility and release profile of clonazepam using simplex lattice design. Journal of Drug Delivery Science and Technology. 2016;**33**:114-124. DOI: 10.1016/j.jddst.2016.04.003

[77] Akhter MH et al. Formulation and development of CoQ10-loaded s-SNEDDS for enhancement of oral bioavailability. Journal of Pharmaceutical Innovation. 2014;**9**(2):121-131. DOI: 10.1007/s12247-014-9179-0

[78] Patel J et al. Formulation and development of self-nanoemulsifying granules of olmesartan medoxomil for bioavailability enhancement. Particulate Science and Technology. 2014;**32**(3):274-290. DOI: 10.1080/02726351.2013.855686

[79] Gumaste SG et al. Development of solid SEDDS, IV: Effect of adsorbed lipid and surfactant on tableting properties and surface structures of different silicates. Pharmaceutical Research. 2013;**30**(12):3170-3185. DOI: 10.1007/s11095-013-1114-4

[80] Ito Y et al. Oral solid gentamicin preparation using emulsifier and adsorbent. Journal of Controlled Release. 2005;**105**(1):23-31. DOI: 10.1016/j.jconrel.2005.03.017

[81] Krupa A et al. Preformulation studies on solid self-emulsifying systems in powder form containing magnesium aluminometasilicate as porous carrier. AAPS PharmSciTech. 2015;**16**(3):623-635. DOI: 10.1208/s12249-014-0247-z

[82] Weerapol Y et al. Enhanced dissolution and oral bioavailability of nifedipine by spontaneous emulsifying powders: Effect of solid carriers and dietary state. European Journal of Pharmaceutics and Biopharmaceutics. 2015;**91**:25-34. DOI: 10.1016/j.ejpb.2015.01.011

[83] Chavan RB, Modi SR, Bansal AK. Role of solid carriers in pharmaceutical performance of solid supersaturable SEDDS of celecoxib. International Journal of Pharmaceutics. 2015;**495**(1):374-384. DOI: 10.1016/j.ijpharm.2015.09.011

[84] Čerpnjak K et al. Tablets and minitablets prepared from spray-dried SMEDDS containing naproxen. International Journal of Pharmaceutics. 2015;**495**(1):336-346. DOI: 10.1016/j.ijpharm.2015.08.099

[85] Li L, Yi T, Lam CW-K. Effects of spray-drying and choice of solid carriers on concentrations of Labrasol® and Transcutol® in solid self-microemulsifying drug delivery systems (SMEDDS). Molecules. 2013;**18**(1):545-560

[86] Tan A, Rao S, Prestidge CA. Transforming lipid-based oral drug delivery systems into solid dosage forms: An overview of solid carriers, physicochemical properties, and biopharmaceutical performance. Pharmaceutical Research. 2013;**30**(12):2993-3017. DOI: 10.1007/s11095-013-1107-3

[87] Wang Z et al. Analysis of DNA methylation status of the promoter of human telomerase reverse transcriptase in gastric carcinogenesis. Archives of Medical Research. 2010;**41**(1):1-6. DOI: 10.1016/j.arcmed.2009.11.001

[88] Newton M et al. The influence of formulation variables on the properties of pellets containing a self-emulsifying mixture. Journal of Pharmaceutical Sciences. 2001;**90**(8):987-995. DOI: 10.1002/jps.1051

[89] Coviello T et al. Polysaccharide hydrogels for modified release formulations. Journal of Controlled Release. 2007;**119**(1):5-24. DOI: 10.1016/j.jconrel.2007.01.004

[90] Zvonar A, Bolko K, Gašperlin M. Microencapsulation of self-microemulsifying systems: Optimization of shell-formation phase and hardening process. International Journal of Pharmaceutics. 2012;**437**(1):294-302. DOI: 10.1016/j.ijpharm.2012.08.013

[91] Franceschinis E et al. Influence of process variables on the properties of simvastatin self-emulsifying granules obtained through high shear wet granulation. Powder Technology. 2015;**274**:173-179. DOI: 10.1016/j.powtec.2015.01.026

[92] Seo A et al. The preparation of agglomerates containing solid dispersions of diazepam by melt agglomeration in a high shear mixer. International Journal of Pharmaceutics. 2003;**259**(1):161-171. DOI: 10.1016/S0378-5173(03)00228-X

[93] Gupta MK et al. Enhanced drug dissolution and bulk properties of solid dispersions granulated with a surface adsorbent. Pharmaceutical Development and Technology. 2001;**6**(4):563-572. DOI: 10.1081/PDT-120000294

[94] Gupta S, Chavhan S, Sawant KK. Self-nanoemulsifying drug delivery system for adefovir dipivoxil: Design, characterization, in vitro and ex vivo evaluation. Colloids and Surfaces A: Physicochemical and Engineering Aspects. 2011;**392**(1):145-155. DOI: 10.1016/j.colsurfa.2011.09.048

[95] Agrawal AG, Kumar A, Gide PS. Formulation of solid self-nanoemulsifying drug delivery systems using N-methyl pyrrolidone as cosolvent. Drug Development and Industrial Pharmacy. 2015;**41**(4):594-604. DOI: 10.3109/03639045.2014.886695

[96] Gamal W, Fahmy RH, Mohamed MI. Development of novel amisulpride-loaded solid self-nanoemulsifying tablets: Preparation and pharmacokinetic evaluation in rabbits. Drug Development and Industrial Pharmacy. 2017;**43**(9):1539-1547. DOI: 10.1080/03639045.2017.1322608

[97] Velasco MV et al. Flow studies on maltodextrins as directly

compressible vehicles. Drug Development and Industrial Pharmacy. 1995;**21**(10):1235-1243. DOI: 10.3109/03639049509026672

[98] Pandey P et al. Formulation and evaluation of herbal effervescent granules incorporated with Martynia Annua extract. Journal of Drug Discovery and Therapeutics. 2013;**1**(5):54-57

Chapter 5

Nanofibrous Scaffolds for Skin Tissue Engineering and Wound Healing Based on Nature-Derived Polymers

Lucie Bacakova, Julia Pajorova, Marketa Zikmundova, Elena Filova, Petr Mikes, Vera Jencova, Eva Kuzelova Kostakova and Alla Sinica

Abstract

Nanofibrous scaffolds belong to the most suitable materials for tissue engineering, because they mimic the fibrous component of the natural extracellular matrix. This chapter is focused on the application of nanofibers in skin tissue engineering and wound healing, because the skin is the largest and vitally important organ in the human body. Nanofibrous meshes can serve as substrates for adhesion, growth and differentiation of skin and stem cells, and also as an antimicrobial and moisture-retaining barrier. These meshes have been prepared from a wide range of synthetic and nature-derived polymers. This chapter is focused on the use of nature-derived polymers. These polymers have good or limited degradability in the human tissues, which depends on their origin and on the presence of appropriate enzymes in the human tissues. Non-degradable and less-degradable polymers are usually produced in bacteria, fungi, algae, plants or insects, and include, for example, cellulose, dextran, pullulan, alginate, pectin and silk fibroin. Well-degradable polymers are usually components of the extracellular matrix in the human body or at least in other vertebrates, and include collagen, elastin, keratin and hyaluronic acid, although some polymers produced by non-vertebrate organisms, such as chitosan or poly(3-hydroxybutyrate-*co*-3-hydroxyvalerate), are also degradable in the human body.

Keywords: skin replacements, wound dressings, nanofibers, electrospinning, epidermis, dermis, keratinocytes, fibroblasts, stem cells, vascularization, cell delivery, drug delivery, regenerative medicine

1. Introduction

Nanofibrous scaffolds are one of the most promising materials for skin tissue engineering and wound dressing, because they resemble nanoarchitecture of the native extracellular matrix (for a review, see [1]). Therefore, they can serve as suitable carriers of cells for tissue engineering and also as suitable wound dressings, which are able to protect the wound from external harmful effects, mainly microbial infection, and at the same time, they can keep appropriate moisture and gas exchange at the wound site.

Nanofibrous scaffolds for skin tissue engineering have been fabricated from a wide range of synthetic and nature-derived polymers, which can be either biostable or degradable within the human body. Biostable synthetic polymers used in nanofiber-based skin regenerative therapies include, for example, polyurethane [2], polydimethylsiloxane [3], polyethylene terephthalate [4], polyethersulfone [5], and also hydrogels such as poly(acrylic acid) (PAA, [6]), poly(methyl methacrylate) (PMMA, [7]), and poly[di(ethylene glycol) methyl ether methacrylate] (PDEGMA, [8]). Degradable synthetic polymers typically include poly(ε-caprolactone) (PCL, [9]) and its copolymers with polylactides (PLCL, [10]), polylactides (PLA, [11]) and their copolymers with polyglycolides (PLGA, [12]), and also so-called auxiliary polymers, such as poly(ethylene glycol) (PEG), poly(ethylene oxide) (PEO, [13]) or poly(vinyl alcohol) (PVA, [14]), which facilitated the electrospinning process and improved the mechanical properties and wettability of the chief polymer. However, the synthetic polymers, although they are well-chemically defined and tailorable, are often bioinert, hydrophobic and thus not promoting cell adhesion, and also not well-adhering to the wound site. Therefore, they need to be combined with other bioactive substances, particularly nature-derived polymers.

This chapter is focused on nature-derived polymers used for fabrication of nanofibrous scaffolds for skin tissue engineering and wound healing. The advantages of most of these polymers are their better bioactivity, flexibility, wettability, and adhesion to the wound site. Similarly as synthetic polymers, also nature-derived polymers can be divided into polymers with none or limited degradability, when implanted into human tissues, and polymers well-degradable in human tissues. The first group includes glucans, such as cellulose, schizophyllan, dextran, starch, and other polysaccharides and proteins, such as pullulan, xylan, alginate, pectin, gum tragacanth, gum arabic, silk fibroin, and sericin. The second group of polymers degradable in human tissues includes collagen and its derivative gelatin, elastin, keratin, glycosaminoglycans such as hyaluronic acid, heparin and chondroitin sulfate, and also polymers not produced in the human body, namely chitosan, gellan gum, zein, and poly(3-hydroxybutyrate-*co*-3-hydroxyvalerate) (PHBV).

Some of the polymers degradable in human tissues, such as collagen, gelatin, elastin, keratin, and glycosaminoglycans, contain specific cell-binding motifs in their molecules, for example, specific amino acid sequences in proteins and oligosaccharide domains in glycosaminoglycans, which are recognized by cell adhesion receptors of integrin and non-integrin families (for a review, see [15, 16]). These molecules are often used in allogeneic or xenogeneic form, thus they can be associated with pathogen transmission or immune reaction. However, some synthetic polymers, for example PLA and PCL, have been reported to induce a more pronounced inflammatory reaction than gelatin [17].

This review chapter summarizes earlier and recent knowledge on skin tissue engineering and wound dressing applications, based on nanofibrous scaffolds made of nature-derived polymers, including our results.

2. Nature-derived nanofibers with none or limited degradability in the human tissues

Nature-derived nondegradable polymers or polymers with limited degradability in human tissues include polymers not occurring in the human body and synthesized by other organisms, such as plants, algae, fungi, insects, and bacteria.

Cellulose is a typical natural polymer nondegradable in human tissues. Cellulose belongs to the group of glucans, that is, polysaccharides derived from D-glucose, linked by glycosidic bond. In the cellulose molecules, these glycosidic bonds are of the β-type, thus the cellulose is a β-glucan. It is structural polysaccharide consisting of a linear chain of several hundred to over ten thousand β(1 → 4) linked D-glucose units. Cellulose is synthesized by plants, algae, fungi, some species of bacteria (*Gluconacetobacter xylinus*), and also by some animals, namely tunicates (*Styela clava*) (for a review, see [18, 19]).

Nanofibrous cellulose can be prepared in three basic forms: bacterial cellulose, which contains cellulose nanofibrils, synthesized by bacteria, nanofibrillar cellulose prepared from plants, particularly from wood, by hydrolysis, oxidation, and mechanical disintegration, and cellulose nanofibers created by electrospinning (for a review, see [19]). For electrospinning, cellulose should be solved. Well-known solvent of cellulose is N-methylmorpholine-N-oxide (NMMO). Another possibility is N-alkylinidazolium-derivate ionic liquid and N,N-dimethylacetamide containing 8 wt% of LiCl. However, any of them did not prove to be a good solvent for needleless electrospinning. The most favorable solvent of cellulose was found to be trifluoroacetic acid (TFA). However, TFA causes severe skin burns and is toxic for aquatic organisms even in low concentrations [20]. These problems, which limit the use of cellulose for creation of electrospun scaffolds for biomedical applications, can be solved by substituting the natural cellulose by its derivatives. The mostly used derivative of cellulose is cellulose acetate (CA), mainly due to its easier solubility and biocompatibility. CA can be dissolved in several solvents, however the best ones for electrospinning proved to be acetic acid (AA), and mixtures of acetone and N,N-dimethylacetamide (DMAC). Some results of successfully spun fibers by needleless electrospinning in our experiments can be found in **Figure 1**, demonstrating differences in the fiber morphology. The 95% aqueous mixture of AA showed the best results in comparison with acetone/DMAC mixtures due to production of smoother fibers and lower cytotoxicity.

All the mentioned forms of cellulose have been widely applied as wound dressings releasing various bioactive agents into wounds (antimicrobial, anti-inflammatory, antioxidative agents, cytokines, and growth and angiogenic factors), as transparent wound dressings for direct optical monitoring of wounds, for systemic

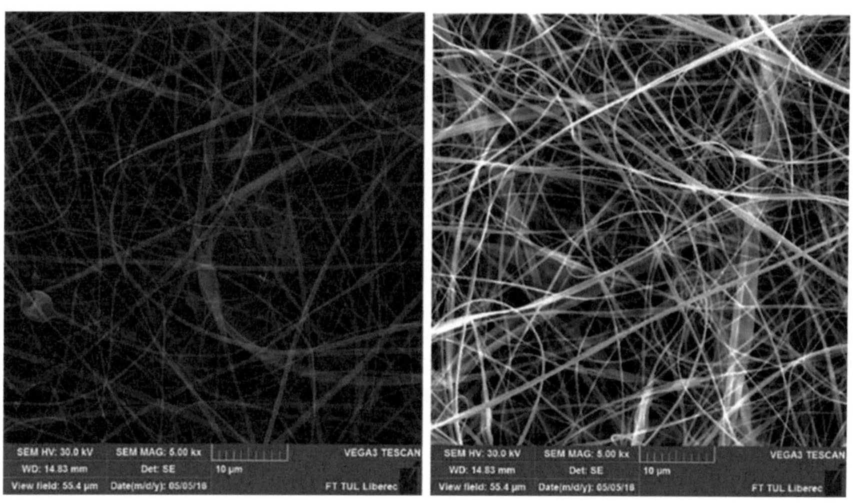

Figure 1.
Scanning electron microscopy of nanofibrous layers produced by wire needleless electrospinning using different solvents, namely 12 wt% of CA in acetone/DMAC (9:1) (left) or 14 wt% of CA in 95% AA (right).

transdermal drug delivery (analgesics, antiphlogistics, corticoids, and antihypertensives) and for construction of epidermal electronics for monitoring wound healing or physiological status of the organism. Non-degradable nanocellulose has also been used as a temporary carrier for delivery of keratinocytes, dermal fibroblasts, and mesenchymal stem cells into wounds (for a review, see [19]).

However, for use as direct scaffolds for skin tissue engineering, cellulose should be rendered degradable in human tissues. Cellulose is degradable by cellulase enzymes (exoglucanases and endoglucanases), which hydrolyze 1,4-beta-D-glycosidic linkages. These enzymes are not synthesized in human tissues, but they can be incorporated into cellulose scaffolds in order to degrade them gradually [21, 22]. These enzymes are believed to be non-toxic for mammalian cells [23, 24]. Moreover, the final product of cellulose degradation by these enzymes is glucose, which is a natural nutrient for the cells, by contrast with the acidic by-products of the standard currently used biodegradable PLA or PLGA scaffolds [25]. Another possibility how to use cellulase enzymes in skin tissue engineering (and in tissue engineering in general) is cell sheet technology. First, cells can be grown on the top of non-degradable cellulose substrates. After reaching the cell confluence, self-standing cell sheets can be released by exposure of the cellulose substrates to cellulases. Unlike the proteolytic enzymes conventionally used for detaching cells from their growth supports, cellulases do not disintegrate the extracellular matrix (ECM) formed by cells and do not cleave extracellular parts of cell adhesion receptors binding the ECM [26]. The cell sheets can be then replanted in the wound bed.

Another interesting approach how to render the cellulose degradable was metabolic engineering of *Gluconacetobacter xylinus*, which then produced modified cellulose molecules with intercalated N-acetylglucosamine (GlcNAc) residues, susceptible to degradation with lysozyme, present in the human body. After subcutaneous implantation in mice, the modified cellulose was completely degraded within 20 days [27, 28].

Other approaches how to render the cellulose degradable, at least partially, is its oxidation and other chemical modifications of cellulose, such as its conversion into regenerated cellulose or 2,3-dialdehydecellulose. In addition, cellulose of animal origin, that is, from tunicates, degraded more quickly than plant cellulose. For example, when cellulose films from *Styela clava* were implanted subcutaneously into rats for 90 days, they lost almost 24% of their initial weight, while the films prepared from wood pulp cellulose lost only less than 10% (for a review, see [19]).

Schizophyllan is another β-glucan used for potential wound healing application. It is an extracellular β-1,3 beta-glucan with β-1,6 branching, produced by the fungus *Schizophyllum commune*. In blends with PVA, it was used for electrospinning of nanofibrous scaffolds, which provided a suitable growth support for human dermal fibroblast. In experimental wound models *in vivo*, schizophyllan attracted macrophages, necessary for the first physiological phase of wound healing, that is, inflammation. Schizophyllan and other 1,3-β-glucans also increased collagen deposition, cellularity, formation of granulation tissue, and vascularity at the wound site [29].

Other glucans used for fabrication of nanofibrous scaffolds for skin tissue engineering and wound healing include dextran, starch and pullulan. According to the type of their glycosidic bonds, these polysaccharides belong to α-glucans.

Dextran is a branched complex glucan, in which the D-glucose units are linked by α-1,6 glycosidic bonds with branches from α-1,3 linkages. Dextran is of microbial origin; it can be produced, for example, by some lactic acid bacteria from sucrose. Dextran was used as a component of nanofibrous polyurethane-based wound dressings, in which dextran promoted neovascularization of the wound site, and also served as carrier for β-estradiol, an endogenous estrogen, a potent anti-inflammatory agent, and mitogen for keratinocytes. In addition, the presence of dextran made the polyurethane dressing softer, more flexible, more wettable, and well-adherent to the wound and promoted hemostatic activity of the dressing. *In vitro*, the presence of dextran and β-estradiol enhanced the proliferation of 3T3-L1 fibroblasts on the scaffolds [30].

Dextran was also used as component of a bilayer scaffold for skin tissue engineering. The upper part of the scaffolds was made of electrospun blend of poly (ε-caprolactone-*co*-lactide) and poloxamer (i.e., Pluronic), and the lower part was made of a hydrogel composed of dextran and gelatin without the addition of a chemical crosslinking agent. The lower dextran/gelatin hydrogel layer provided a highly swollen three-dimensional environment similar to extracellular matrix (ECM) of soft tissues. Both part of the scaffolds supported the growth of adiposetissue-derived stem cells; however, the number of these cells on the hydrogels decreased with increasing content of dextran [31].

Dextran is degradable by dextranases, enzymes hydrolyzing (1 → 6)-alpha-D-glycosidic linkages. This enzyme is produced mainly by bacterial and fungi, but it was also detected in animal and human tissues, namely liver and spleen. Therefore, dextran is often chosen for biomedical applications, particularly drug delivery, because it is slowly degradable in human organism. Dextran molecules with Mw higher than 40 kDa are sequestered in the liver and spleen, and then hydrolyzed by endo- and exodextranases. Dextran molecules with Mw lower than 40 kDa can be eliminated through renal clearance [32]. However, dextran hydrogels implanted subcutaneously or intramuscularly into rats did not show signs of degradation 6 weeks post-implantation and were surrounded by a thin fibrous capsule and some macrophages and giant cells, which is a response typical for a number of non-degradable materials [32].

Starch is another α-glucan, containing both α-1,4- and α-1,6 glycosidic bonds. It serves an energy storage polysaccharide in plants, and from this point of view, it is considered to be an analogue of glycogen, energy storage polysaccharide in animals. Starch consists of two types of molecules, namely linear amylose and branched amylopectin (for a review, see [33]). Electrospun starch-based nanofibrous meshes were proposed for wound healing applications. The electrospinning of starch was facilitated by addition of PVA, that is, a noncytotoxic, water-soluble, biocompatible synthetic polymer which reduced the repulsive forces produced in starch solution. The scaffolds then promoted the proliferation of mouse L929 fibroblasts [34]. Starch is degradable by amylases, that is, hydrolases that act on α-1,4-glycosidic bonds. Amylases occur in three forms, namely α-, β-, and γ-amylases. These enzymes are synthesized by microorganisms (bacteria and fungi), plants, and with exception of β-amylases, also in animals. Alpha-amylases are present in human organism, but not currently in all tissues-they are important enzymes of gastrointestinal tract and are produced by salivary glands and pancreas. Interestingly, α-amylases were also found in brain, and their lower expression there is probably associated with the pathogenesis of Alzheimer's neurodegenerative disease [35].

Pullulan is also an α-glucan with both α-1,4- and α-1,6 glycosidic bonds. It is a linear polysaccharide consisting of maltotriose units, in which three glucose units in maltotriose are connected by an α-1,4 glycosidic bond, whereas consecutive maltotriose units are connected to each other by an α-1,6 glycosidic bond. Pullulan is produced from starch by the fungus *Aureobasidium pullulans*. It shows a high water-absorbing capability, adhesive properties, and the capability to form strong resilient films and fibers. It is degradable by pullulanase, a specific kind of glucanase, produced in bacteria and not present in human tissues. When pullulan hydrogels alone or in combination with dextran were implanted subcutaneously into rats, they induced inflammatory reaction and were surrounded by a fibrous capsule [36]. Nevertheless, pullulan is water-soluble and thus removable from human issues, and in combination with chitosan and tannic acid, it was used for fabrication of electrospun nanofibrous meshes promising for wound healing [37]. In combination with dextran and gelatin, pullulan was used for electrospinning of nanofibrous scaffolds promising for skin tissue engineering. These scaffolds, especially when crosslinked with trisodium trimetaphosphate, supported the adhesion and spreading of human dermal fibroblasts and formation of actin cytoskeleton in these cells [38].

Xylan is a plant polysaccharide belonging to the group of hemicelluloses, that is, polymers often associated with cellulose. While cellulose is made of glucose units, hemicelluloses contain many different sugar monomers. Xylans are polysaccharides made of β-1,4-linked xylose (i.e., a pentose sugar) residues with side branches of α-arabinofuranose and α-glucuronic acids, which contribute to crosslinking of cellulose microfibrils and lignin through ferulic acid residues. Xylans are considered as relatively available and cost-effective natural materials for tissue engineering. Electrospun nanofibers containing beech-derived xylan and PVA were tested as potential dermal substitutes for skin tissue regeneration. These scaffolds provided a good support for the adhesion and proliferation of human foreskin fibroblasts and for production of collagen by these cells [39]. Bagasse xylan was also a component of hydrogels endowed with shape memory, namely carboxymethyl xylan-g-poly(acrylic acid) hydrogels, applicable in tissue engineering and biosensorics, particularly for construction of electronic skin [40].

Alginates, for example, sodium alginate or calcium alginate, are salts of alginic acid, a linear polysaccharide composed of (1,4)-β-D-mannuronic acid and (1,3)-α-L-guluronic acid. Alginates are produced by various species of brown algae, and also by the bacterium *Pseudomonas aeruginosa*, a major pathogen found in the lungs of patients with cystic fibrosis. The structure of alginates is similar to glycosaminoglycans, an important component of ECM in human tissues including skin [41]. Alginates have a great ability to keep moisture in the wound site and to adhere to skin. However, alginates are poorly spinnable, and therefore, for skin tissue engineering and wound dressing applications, they were electrospun together with other polymers, such as PVA [41, 42] or PEO [43]. Poor mechanical properties of alginates have been compensated by the combination with chitosan [44] or PCL [44, 45]. In addition, alginates themselves are not adhesive for mammalian cells, which was compensated by their combination with collagen and gelatin, containing ligands for cell adhesion receptors [41]. Alginates were modified with a cell adhesive GRGDSP oligopeptide, which acts as ligand for integrin cell adhesion receptors [43]. Sodium alginate was used for attachment of arginine to the surface of chitosan nanofibers in order to increase healing capability of this wound dressing [46]. Alginate nanofibers supported by PCL were impregnated with an extract from *Spirulina*, a photosynthetic cyanobacterium producing bioactive molecules with anti-oxidant and anti-inflammatory effects [45]. Electrospun sodium alginate nanofibers containing silver nanoparticles were used for fabrication of an electronic skin capable of pressure sensing and endowed with antibacterial activity [47].

The degradability of alginate in human organism is limited. Alginate is naturally degraded by alginate lyases or alginate depolymerases, which have been isolated from marine algae, marine animals, bacteria, fungi, viruses, and other microorganisms, but are not present in the human organism. Degradability of alginate can be increased by its oxidation and at low pH. Also the hydrophilicity and water uptake capacity of alginate can help in its removal from the wound site (for a review, see [48]).

Pectin is a complex of structural polysaccharides present in the cell walls of terrestrial plants, rich in galacturonic acid. Pectin is known as gelling agent in food industry, but it is also widely used in medicine, for example, against digestive disorders, such as obstipation and diarrhea, for oral drug delivery, as a component of dietary fibers trapping cholesterol and carbohydrates, as a demulcent, that is, a mucoprotective agent, and also in wound healing preparations [49]. Pectin is known as a natural prophylactic substance against poisoning with toxic cations, and its hemostatic and curing effects are well-documented in healing ointments [50]. Pectin is degradable by enzymes produced by bacterial, fungal and plant cells, and not present in human tissues [51–53]. Thus, pectin is degradable, at least partly, only in the intestinal tract populated with bacteria. However, pectin is water-soluble and quickly dissolves in the water environment, including the tissues. Therefore, in order to increase its stability, it was combined with chitosan and TiO_2 nanoparticles for wound dressing applications [50] or used for construction of composite chitosan-pectin scaffolds for skin tissue engineering. Blending chitosan with pectin markedly improved the mechanical properties of the scaffolds, such as their Young's modulus, strain at break and ultimate tensile strength, in comparison with pure chitosan scaffolds, although the proliferation of cells (i.e., fibroblasts) was slightly slower on pectin-containing scaffolds [54, 55]. The reason is that pectin does not contain cell binding domains. The cell adhesion on pectin nanofibers was markedly enhanced by oxidizing pectin with periodate to generate aldehyde groups, and then crosslinking the nanofibers with adipic acid dihydrazide to covalently connect pectin macromolecular chains with adipic acid dihydrazone linkers. In addition, the crosslinked pectin nanofibers exhibited excellent mechanical strength and enhanced body degradability [56].

Other polysaccharides explored for creation of nanofibrous scaffolds for skin tissue engineering and wound healing are gum tragacanth and gum arabic, both polysaccharides of plant origin, degradable by bacteria and fungi, for example, in soil [57, 58].

Gum tragacanth is a viscous water-soluble mixture of polysaccharides, mainly tragacanthin and bassorin. Tragacanthin dissolves to give a colloidal hydrosol. Bassorin, representing 60–70% of the gum, is insoluble and swells to a gel. Chemically, tragacanthin is a complex mixture of acidic polysaccharides containing D-galacturonic acid, D-galactose, L-fucose (6-deoyl-L-galactose), D-xylose, and L-arabinose. Bassorin is probably a methylated tragacanthin. A small amount of cellulose, starch, protein and ash are also present (https://colonygums.com/tragacanth). In order to improve electrospinning and mechanical properties of the gum tragacanth, it was combined with PVA and PCL [59]. Gum tragacanth is endowed with microbial resistance and wound healing activity, which was further enhanced by curcumin, a naturally occurring poly-phenolic compound with a broad range of favorable biological functions, including anti-cancer, anti-oxidant, anti-inflammatory, anti-infective, angiogenic, and healing properties [60].

Gum arabic, also known as gum acacia, is a complex and water-soluble mixture of glycoproteins and polysaccharides consisting mainly of arabinose and galactose. For skin tissue engineering, it was electrospun with PCL and also with zein, a storage plant protein [61].

Silk fibroin is a water-insoluble elastic protein present in silk fibers produced by larvae of *Bombyx mori* and some other moth of the *Saturniidae* family, such as *Antheraea assama, Antheraea mylitta*, and *Philosamia ricini* [62–64]. Silk fibroin occurs in the fibers together with sericin, a water-soluble serine-rich protein, which forms a glue-like layer coating two singular filaments of fibroin.

In biomaterial science, silk fibroin is considered to be degradable, but in mammalian organism, this degradation is long-lasting and can take more than 1 year. As a kind of biomaterial approved by the Food and Drug Administration (FDA) for medical use, silk is defined by United States Pharmacopeia as non-degradable for its negligible tensile strength loss *in vivo*. However, silk fibroin is susceptible to biological degradation by proteolytic enzymes such as chymotrypsin, actinase, carboxylase, proteases XIV, XXI and E, and collagenase IA. The final degradation products of silk fibroin are amino acids, which are easily absorbed *in vivo* (for a review, see [65]).

The degradation behavior of fibroin scaffolds depends on the preparation method and structural characteristics, such as processing condition, pore size, and silk fibroin concentration (for a review, see [65]). For example, three-dimensional porous scaffolds prepared from silk fibroin using all-aqueous process degraded within 2–6 months after implantation into muscle pouches of rats, while the scaffolds prepared using an organic solvent, hexafluoroisopropanol (HFIP), persisted beyond 1 year. It was probably due to a lower original silk fibroin concentration, larger pore size, and a higher and more homogeneous cellular infiltration of aqueous-derived scaffolds than in HFIP-derived scaffolds [66].

For skin tissue engineering and wound healing, silk fibroin has been combined with various synthetic and natural polymers and other bioactive substances. The polymers included, for example, PCL, [67], poly(L-lactic acid)-*co*-poly(ε-caprolactone) (PLACL, [68]), carboxyethyl chitosan, PVA, [69], chitin [70], cellulose-based materials modified by oxidation [71] or with lysozyme [72], collagen [73], gelatin [74], and hyaluronan [75]. The bioactive substances were, for example, growth factors, such as epidermal growth factor [64], vitamins, such as vitamin C [68], vitamin E [76], and pantothenic acid (vitamin B5; [77]), antioxidants, such as grape seed extract ([78]) or quinone-based chromenopyrazole [79], antibiotics, such as ciprofloxacin [64], tetracycline [68] or gentamycin [62], and other antimicrobial and wound healing agents, such as silver nanoparticles, dandelion leaf extract [63], *Aloe vera* [80], or astragaloside IV [74]. In order to enlarge the pore size in nanofibrous scaffolds for cell penetration, silk fibroin was electrospun together with so-called "sacrificial" crystals of ice [67] or NaCl [81, 82], that is, crystals which are removed after the electrospinning process. An interesting combination is silk fibroin with decellularized human amniotic membrane, which was used for developing a three-dimensional bi-layered scaffold for burn treatment. Adipose tissue-derived mesenchymal stem cells seeded on this scaffold increased expression of two main pro-angiogenesis factors, vascular endothelial growth factor, and basic fibroblast growth factor [83]. Also the transplantation of bone marrow-derived mesenchymal stem cells and epidermal stem cells into wounds using nanofibrous silk fibroin scaffolds supported re-epithelization, collagen synthesis, as well as the skin appendages regeneration [84]. Another interesting approach is to use silk fibroin produced by other species than *Bombyx mori*, namely by the moths *Antheraea assama* and *Philosamia ricini*. This "non-mulberry" silk fibroin possesses inherent Arg-Gly-Asp (RGD) motifs in its protein sequence, which facilitates binding of cells through their integrin adhesion receptors [64].

Sericin has also been applied in skin tissue engineering and wound healing, although in a lesser extent than silk fibroin. Sericin shows antioxidant, UV-protective, heat-protective, moisture-retaining, and antimicrobial properties, which have been reported to be more pronounced in non-mulberry sericin (e.g., from *Antheraea mylitta*) than in sericin produced by *Bombyx mori*. The reason is that wild moths like *Antheraea mylitta* are exposed to a hostile environment in nature than *Bombyx mori* raised in captive conditions. Similarly as non-mulberry silk fibroin, also sericin has been reported to be more supportive for cell adhesion than mulberry sericin (for a review, see [85]). Sericin enhanced the proliferation and epidermal differentiation of human mesenchymal stem cells on gelatin/hyaluronan/chondroitin sulfate nanofibrous scaffolds [86]. Similarly, sericin improved the growth of murine L929 fibroblasts and human HaCaT keratinocytes cultured on the PVA nanofibrous scaffolds [87] and also the growth of L929 fibroblasts on chitosan nanofibrous scaffolds, together with antibacterial properties of these scaffolds [88].

3. Nature-derived nanofibers degradable in the human tissues

Nature-derived polymers degradable in human tissues include, in particular, polymers that are synthesized in the human body and usually act as components of ECM. These polymers are proteins (collagen and its derivative gelatin, elastin, fibrinogen and fibrin, keratin) or polysaccharides in non-sulfated form (hyaluronic acid) and sulfated form (heparin-like glycosaminoglycans). In addition, some natural polymers synthesized by other organisms, such as bacteria, fungi, insects, crustaceans or plants, are degradable in human tissues, because they are susceptible to enzymes present in human tissues, such as lysozyme and esterases. These polymers include chitosan, gellan gum, zein, and PHBV.

Collagen is the main structural protein in the extracellular space in a wide range of tissues in the body. Skin contains type I collagen, one of the most abundant collagens in the human body. Type I collagen is also abundant in tendons, ligaments, and vasculature, and it is a main component of the organic part of bone. Type I collagen is a fibrillar type of collagen; it is composed of amino acid chains forming triple-helices of elongated fibrils. That is why the nanofibrous collagen scaffolds closely mimic the architecture of the native ECM and are advantageous for tissue engineering. In addition, collagen has been reported to be relatively poorly immunogenic, even if used in allogeneic and xenogeneic forms, for example, recombinant human collagen or bovine and porcine collagen. However, mammalian collagen is associated with the risk of disease transmission, for example, bovine spongiform encephalopathy (for a review, see [89–91]). This risk can be reduced by the use of fish collagen, which became to be popular in tissue engineering, including skin tissue engineering and wound healing. In addition, the fish collagen enables an easier recovery of intact collagen triple helices than the mammalian collagen [92]. Fish collagen can be obtained from the skin, scales and bones of freshwater fish, such as tilapia [91–94], and marine fish, such as hoki fish (*Macruronus novaezelandiae*) [92, 95], or *Arothron stellatus,* also known as "stellate puffer," "starry puffer" or "starry toadfish" [96]. Nanofiber electrospun from tilapia skin collagen promoted the proliferation of human HaCaT keratinocytes, and stimulated epidermal differentiation through the up-regulated gene expression of involucrin, filaggrin, and type I transglutaminase in these cells. Moreover, the tilapia collagen nanofibers accelerated wound healing *in vivo* in rat models [91–94]. Beneficial effects on wound healing were also observed in nanofibrous meshes electrospun from collagen obtained from *Arothron stellatus* [96] and from fish scale collagen peptides [90].

Collagen is one of the most widely used natural proteins for creation of nanofibrous scaffolds for skin tissue engineering and wound healing. However, these scaffolds are usually mechanically weak, and therefore they need crosslinking or blending with synthetic polymers. Collagen crosslinking with conventionally used agents, particularly glutaraldehyde, is associated with the risk of the scaffold cytotoxicity. More benign crosslinkers used recently include, for example, citric acid [95] or quaternary ammonium organosilane, a multifunctional crosslinking agent, which improved the electrospinnability of collagen by reducing its surface tension, endowed the collagen nanofibers with potent antimicrobial activity and promoted the adhesion and metabolic activity of primary human dermal fibroblasts without any cytotoxicity, at least in a lower concentration of 0.1% w/w [97].

Synthetic polymers used for combination with collagen in nanofibrous scaffolds included PLA [98], PLGA [99, 100], and particularly PCL, which was either blended with collagen [101–104] or served as substrate for subsequent deposition of collagen [105]. Collagen has also been combined with natural polymers, such as silk fibroin [73] or chitosan in a form of blends [106] or in a form of bilayered scaffolds, where collagen was electrospun onto the chitosan scaffolds [107]. Collagen was also grafted on the surface of composite electrospun PVA/gelatin/alginate nanofibers [41]. Collagen-based or collagen-containing nanofibers have been loaded with a wide range of bioactive substances, such as vitamin C, vitamin D3, hydrocortisone, insulin, triiodothyronine, and epidermal growth factor [100], transforming growth factor-β1 [102], plant extracts such as *Coccinia grandis* leaf extract [96], or lithospermi radix extract [107], antibiotics such as gentamicin [103], or bioactive glass [93, 104]. Collagen and PCL and bioactive glass nanoparticles were applied for delivery of endothelial progenitor cells into wounds in order to promote their vascularization and healing [104].

Gelatin is a derivative of collagen, obtained by denaturing its triple helical structure. Specifically, it is a mixture of peptides and proteins produced by partial hydrolysis of collagen extracted from the skin, bone, and connective tissue of animals, such as cattle, pigs, chicken, and also fish. Gelatin can be defined as a complex mixtures of oligomers of the α subunits joined by covalent bonds, and intact and partially hydrolyzed α-chains of varying molecular weight (for a review, see [89, 92]). Properties of gelatin, including its spinnability, depend on the source of collagen, animal species, age, type of collagen, type of conversion of collagen to gelatin (i.e., acidic *vs.* basic hydrolysis), and particularly on the conformation of gelatin molecules [108].

Similarly as collagen, also gelatin is the most promising for skin tissue engineering and wound healing applications in combination with various synthetic and natural polymers. For example, gelatin was combined with polyurethane [109], PLA [11, 17], and particularly with PCL, where it was incorporated into core-shell PCL/gelatin nanofibers as the core polymer [110] or electrospun independently of PCL using a double-nozzle technique, which resulted in creation of two types of nanofibers in the scaffolds, either mixed [111] or arranged in separate gelatin and PCL layers [112]. Gelatin was also combined with a copolymer of lactic acid and caprolactone P(LLA-CL) in the form of blends [113] or in the form of coaxial nanofibers with P(LLA-CL)/gelatin shell and albumin core containing epidermal growth factor, insulin, hydrocortisone, and retinoic acid [114]. Natural proteins combined with gelatin included dextran [31], pullulan [38], alginate [41], silk fibroin [74], and hyaluronan with chondroitin sulfate [86].

For combination with synthetic and natural polymers, for example, with PCL [115], and chitosan and keratin [116], gelatin was also used in the form of photo-crosslinkable gelatin methacrylate hydrogel (GelMA). On PCL nanofibers, GelMA showed a higher decoration level in comparison with native gelatin [116]. Self-standing nanofibrous matrices electrospun from GelMA enabled tuning of their water retention capacity, stiffness, strength, elasticity, and degradation by changing the exposure time to UV light [117].

Elastin is a protein found in the ECM, that maintains its elasticity [118]. It is the second main protein-based component of native skin ECM. The presence of elastin in composite electrospun nanofibrous scaffolds, containing gelatin, cellulose acetate and elastin, changed the fiber morphology from straight to ribbon-like structure, and decreased the swelling ratio and degradation rate of the scaffolds. In addition, elastin-containing scaffolds supported the attachment and proliferation of human fibroblasts [119].

Fibrin is a provisional ECM protein, which accumulates in wounds after injury to initiate hemostasis and healing. Fibrin is formed via the polymerization of fibrinogen monomers in the presence of thrombin, and this process can be simulated *in vitro* [120]. Fibrin forms a fine nanofibrous mesh, which is mechanically weak and needs to be deposited on some supportive structure, for example, synthetic nanofibrous meshes made of poly(L-lactide) (PLLA) [121]. In our experiments, fibrin was deposited on PLLA in the form of two types of coating, depending on the mode of

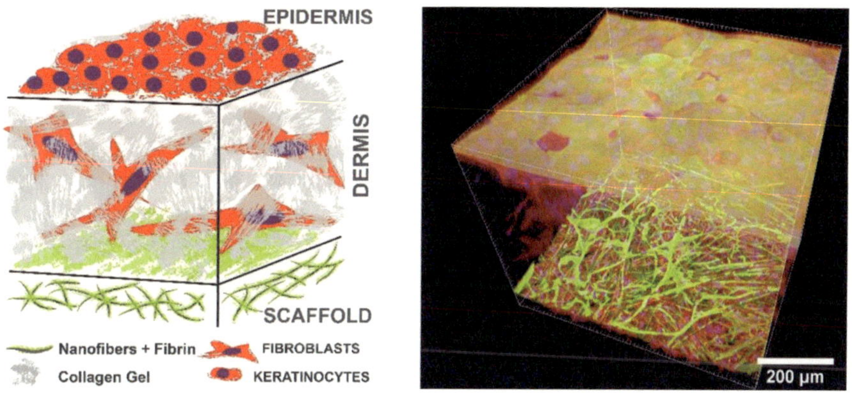

Figure 2.
Developing a bilayer construct of keratinocytes and fibroblasts on a PLLA nanofibrous membrane with fibrin and collagen hydrogel. Left: schematic design; right: real construct.

fibrin preparation. Fibrin either covered the individual fibers in the membrane (F1 nanocoating), or covered the individual fibers and also formed a fine homogeneous nanofibrous mesh on the surface of the membrane (F2 nanocoating), depending on the mode of fibrin preparation. The fibroblasts on the F1 nanocoating remained in their typical spindle-like shape, while the cells on the F2 nanocoating were polygonal with a higher proliferation rate [122]. F2 nanocoating was then used for development of a bilayer skin construct. First, a nanofibrous PLLA mesh was coated with fibrin and seeded with human dermal fibroblasts. After reaching confluence, the fibroblasts were covered with a collagen hydrogel and were allowed to migrate into this hydrogel and to proliferate inside. After sufficient colonization of the hydrogel with fibroblasts and formation of a structure resembling the skin dermis, human epidermal keratinocytes were seeded on the top of the collagen hydrogel (**Figure 2**) [123].

Also fibrinogen was used for modification of synthetic polymeric nanofibers in order to enhance the cell adhesion and growth. Nanofibrous scaffolds electrospun from blends of PCL and fibrinogen improved the adhesion, proliferation, and epidermal differentiation of adipose tissue-derived stem cells (ADSCs) in comparison with pure PCL scaffolds. Composite PCL/fibrinogen scaffolds seeded with ADSCs also markedly improved healing of full-thickness excisional wounds created in rats in comparison with acellular dermal matrix or acellular dermal matrix with ADSCs [124].

Keratin is a fibrous structural protein, present in skin appendages, such as hair, wool, feather, nails, horns, claws, hooves, and in the outer (cornified) layer of epidermis [125, 126]. Keratin protects epithelial cells from damage and stress and is insoluble in water and organic solvents.

In most studies dealing with keratin-containing nanofibers, keratin was combined with other natural or synthetic polymers in order to improve the spinnability of keratin, or to improve the bioactivity of the co-electrospun polymer. For example, in a study by Cruz-Maya *et al.* [127], blending keratin with PCL improved the stability of the electrospinning process, promoted the formation of nanofibers without defects, such as beads and ribbons, typically observed in the fabrication of keratin nanofibers. At the same time, keratin markedly increases the fiber hydrophilicity compared with pure PCL, which improved the adhesion and proliferation of human mesenchymal stem cells [127]. Similarly, co-electrospinning of keratose (i.e., oxidative keratin) with PVA resulted in nanofibers with uniform fibrous structure, suitable hydrophilicity and mechanical properties [125]. Properties of electrospun keratin nanofibers were also improved by incorporation of hydrotalcites, intended for delivery of diclofenac. These nanofibers displayed a reduced swelling ratio and a slower degradation profile compared to keratin-based non-woven nanofibrous mats containing free diclofenac [126]. Keratin was also a component of core-shell nanofibers, prepared by coaxial electrospinning of chitosan, PCL and keratin with *Aloe vera* extracts encapsulated inside the polymer nanofibers. This construct increased the adhesion and growth of L929 fibroblasts and was intended for wound healing applications [128]. Importantly, keratin was a component of bilayer scaffolds for skin tissue engineering, composed of human hair keratin/chitosan nanofiber mat and gelatin methacrylate (GelMA) hydrogel. Human dermal fibroblasts were encapsulated and grown in the hydrogel matrix, while human HaCaT keratinocytes formed a layer on the top of the scaffolds, mimicking dermis and epidermis of skin tissue [116]. Another bilayer scaffolds was constructed using polyurethane wound dressing as an outer layer, and electrospun gelatin/keratin nanofibrous mat as an inner layer [109].

Hyaluronic acid, also called hyaluronan, is an anionic, non-sulfated linear glycosaminoglycan. It is distributed widely throughout connective, neural, and epithelial tissues, including skin, where it is a major component of ECM. Therefore, hyaluronic acid has been widely used for skin tissue engineering and wound healing, and it is approved for clinical application [33].

Hyaluronic acid stimulated infiltration of nanofibrous scaffold composed of hyaluronan, silk fibroin and PCL [75], and can help to promote cell proliferation [129]. Electrospinning of pure hyaluronic acid is not simple because of solubility characteristics of this polymer. Hyaluronic acid is well-soluble in water but less-soluble in most organic solvents, which can be solved by mixtures of solvents as water/ethanol or water/dimethylformamide [130]. Increasing of evaporation and decreasing of solution surface tension by the solvent mixing helps to electrospinning process. Another possibility is electrospinning of hyaluronic acid together with a suitable water-soluble polymer such as PVA [131] or PEO. The solution of pure hyaluronic acid [132] or with relatively small amount of carrier PEO was successfully spun into nanofibrous material by air-assisted electrospinning technology, that is, electroblowing [133]. For creation of nanofibrous scaffolds, hyaluronic acid was also used in combination with PCL [134], PLA [135] or gelatin, chondroitin sulfate and sericin [86].

Sulfated glycosaminoglycans have been relatively rarely used as components of nanofibers for skin tissue engineering and wound healing in comparison with hyaluronic acid. This group of polysaccharides include heparin, heparan sulfate, chondroitin sulfate, dermatan sulfate, and keratan sulfate (for a review, see [33]). From these polysaccharides, only heparin, heparan sulfate, and chondroitin sulfate were used as components of nanofibers for skin regenerative therapies. For example, heparin coatings on PLLA nanofibers increased the infiltration of the scaffolds with endothelial cells *in vitro,* and enhanced epidermal skin cell migration across the wound in a full-thickness dermal wound model in rats *in vivo* [136]. In a recent study by Yergoz *et al*. [137], a heparin-like nanofibrous hydrogel promoted regeneration of full thickness burn injury in mice. Chondroitin sulfate was used as a component of electrospun gelatin/PVA/chondroitin sulfate nanofibrous scaffolds, which supported the proliferation of human dermal fibroblasts [138], of electrospun nanofibrous composite scaffolds made of cationic gelatin/hyaluronan/chondroitin sulfate loaded with sericin, which promoted the differentiation of human mesenchymal stem cells toward epithelial lineage [86], and of electrospun gelatin/chondroitin sulfate nanofibrous scaffolds, which accelerated healing of full-thickness skin excision wounds in rats [139].

Chitosan is a linear polysaccharide composed of randomly distributed β-(1 → 4)-linked D-glucosamine (deacetylated unit) and N-acetyl-D-glucosamine

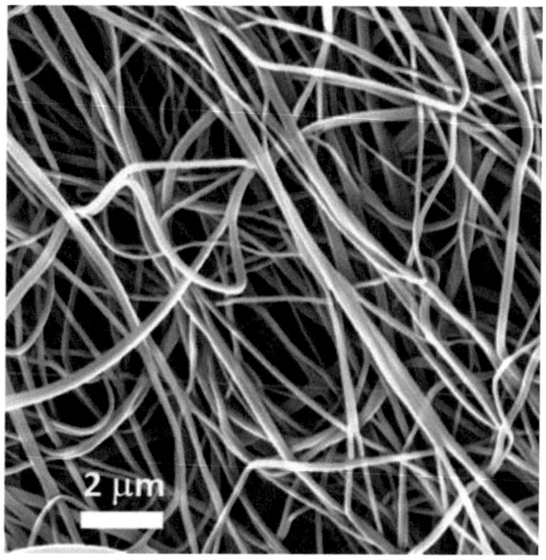

Figure 3.
Scanning electron microscopy of nanofibrous layers produced by needle electrospinning from PVA/chitosan solution.

(acetylated unit). It can be prepared by alkali treatment from chitin, a poly-N-acetyl-D-glucosamine polysaccharide, the major structural component of the exoskeleton of crustaceans (crab, shrimp), and of the cell wall of fungi and yeast [140]. Chitosan is known as biocompatible, antimicrobial, and biodegradable. In the human organism, it can be degraded by lysozyme, a hydrolytic enzyme present in various secretions such as saliva, tears, mucus, and human milk, and also in cytoplasmic granules of macrophages and polymorphonuclear neutrophils. Chitosan breakdown by lysozymes happens via the removal of glycosidic bonds between polysaccharide units in the polymer. Glucosamine and saccharide are the products of this process, which can be metabolized or stored as proteoglycans in the body.

However, electrospinning of chitosan is difficult due to its polycationic characters. Due to the presence of amine groups in the chitosan molecule, acidic aqueous solutions are the best solvents for this polymer. The best candidates for solvent system seem to be mixture of acetic acid (AA) and formic acid (FA) or trifluoroacetic acid (TFA); however, TFA is highly toxic. Electrospinning of pure chitosan has very low productivity because it requires very concentrated polymeric solutions [141]. Therefore, for creation of nanofibrous scaffolds for skin tissue engineering, chitosan has been mixed with other natural or synthetic polymers, such as collagen [142], gelatin [143], keratin [116], cellulose [144], pectin [54, 55], silk fibroin [69], PHBV [145], PCL [142], PLA [146], PLGA [147], PEO, [148], and also with PVA, which was used in our studies (**Figure 3**). Chitosan has also been mixed with various nanoparticles, such as halloysite nanotubes [149], graphene oxide [150] or nanodiamonds [144]. The reason of all these mixtures was to improve the stability, spinnability, wettability, mechanical properties, and biofunctionality of chitosan-containing scaffolds for skin tissue engineering. Combination of chitosan with various polymers also enabled creation of bilayer scaffolds for reconstruction of two main skin layers, that is, epidermis containing keratinocytes and dermis containing fibroblasts [116, 142]. In order to enhance the antimicrobial and wound healing activity of chitosan, this polymer was electrospun together with extract from Henna leaves [151]. In addition, chitosan nanoparticles have been incorporated in nanofibrous scaffolds as carriers for controlled drug delivery, for example, delivery of growth factors, cytokines and angiogenic factors, such as platelet-derived growth factor [152], granulocyte colony-stimulating factor [153] or angiogenin [147].

Nanofibrous scaffolds promising for skin tissue engineering and wound healing were also prepared directly from chitin, which was electrospun either alone with further modifications with fibronectin, laminin and particularly with type I collagen [154], or in combination with silk fibroin [70].

Gellan gum is a water-soluble anionic polysaccharide produced by the bacterium *Sphingomonas elodea* (formerly *Pseudomonas elodea*). The repeating unit of the polymer is a tetrasaccharide, which consists of two residues of D-glucose, one of residue of L-rhamnose and one residue of D-glucuronic acid. For skin tissue engineering and wound healing, gellan gum was electrospun with PVA in order to decrease its viscosity and repulsive forces between the polyanions along the polymer chains and to increase the stability, uniformity, and structural consistency of the nanofibers in aqueous environment. The nanofibrous scaffolds were further stabilized by crosslinking with various physical, chemical, and ionic methods, such as by heat, UV irradiation, methanol, glutaraldehyde, and by calcium chloride [155]. These scaffolds supported the adhesion and growth of human dermal 3T3-L1 fibroblasts [155, 156] and human HaCaT keratinocytes [157] and provided a better support for these cells than conventionally proposed gellan-based hydrogels and dry films. In addition, these scaffolds were endowed with antimicrobial activity by incorporation with amoxicillin, and accelerated healing of full-thickness skin excision wound in rats in comparison with non-treated wounds [157]. Similarly as chitosan, gellan gum has been reported to be degradable by lysozyme [158]. Three-dimensional printed gellan gum scaffolds also showed degradation *in vitro* in phosphate-buffered saline (PBS) or in simulated body fluid, and the degradation rate could be modulated by changing the ratio of surface area per mass of the scaffolds [159].

Zein is the major storage protein of corn, composed of amino acids such as leucine, glutamic acid, alanine and proline, and showing good biocompatibility, flexibility, microbial resistance, and antioxidant activity [61]. Zein has been shown to degradable *in vitro* in PBS and also *in vivo* when implanted subcutaneously in rats in the form of rod-like implants [160]. However, similarly as in many other natural polymers, the application of pure zein nanofibers is limited because of poor mechanical properties of these fibers. Therefore, for skin tissue engineering and wound dressing applications, zein has been mixed with various synthetic and nature-derived polymers, such as polyurethane [161], PLA [162], PCL, hyaluronic acid, chitosan [163], and polydopamine [164], and impregnated with TiO_2 nanoparticles [164] or Ag nanoparticles [161] in order to enhance the antimicrobial activity of the scaffolds.

Poly(3-hydroxybutyrate-co-3-hydroxyvalerate) (PHBV), is a polymer produced naturally by bacteria as a storage compound under growth-limiting conditions. It is a thermoplastic linear aliphatic polyester of polyhydroxyalkanoate type. PHBV is approved by the FDA for medical use. PHBV is biodegradable by bacterial enzymes, but it is also susceptible to hydrolytic degradation in water environment, although this degradation is relatively slow. When degradation of porous PHBV scaffolds for tissue engineering was simulated *in vitro* in PBS at 37°C, it lasted several months [165]. In the human body, however, the degradation of PHBV can be accelerated by nonspecific esterase and lysozyme enzymes, both present in cells of the immune system (for a review, see [165]). For biomedical application, PHBV is often used as an alternative to synthetic polymers, but it has several drawbacks, such as relatively high cost, brittleness, relatively difficult processing, and also hydrophobicity, which can hamper the cell adhesion and growth. However, PHBV is piezoelectric, which can stimulate the adhesion, growth, and phenotypic maturation of cells. Pure electrospun PHBV meshes supported the adhesion, growth, and epidermal differentiation of bone marrow mesenchymal stem cells, which was induced by an appropriate composition of cell culture media, containing epidermal growth factor, insulin, 3,3′,5-triiodo-L-thyronine (T3), hydrocortisone, and 1α, 25-dihydroxyvitamin (D3), and manifested by expression of genes for keratin, filaggrin, and involucrin, that is, an early, intermediate and late marker of keratinocyte differentiation, respectively [166]. In order to increase the attractiveness of electrospun PHBV nanofibers for the cell adhesion and growth, they were coated with collagen [167] blended with collagen [168], blended with chitosan [145] or blended with keratin [169]. Collagen-coated PHBV nanofibers alone or seeded with unrestricted somatic stem cells, isolated from umbilical cord, accelerated closure of excision wounds in rats *in vivo* compared to unmodified PHBV nanofibers [167]. Similar wound healing effect was also obtained with PHBV nanofibers blended with keratin [169]. Mechanical properties of PHBV nanofibers were improved by addition of graphene oxide nanoparticles in the electrospinning solution, which also endowed these fibers with and antimicrobial activity [167].

4. Conclusions

Nanofibrous scaffolds made of nature-derived polymers hold a great promise for skin tissue engineering and wound healing. These scaffolds are created from biological matrices, and from this point of view, they resemble the extracellular matrix more closely than synthetic polymers. Some of these polymers, such as collagen, gelatin, elastin, keratin, nonsulfated and sulfated glycosaminoglycans, and also nonmulberry silk fibroin, contain motifs that are recognized and bound by cell adhesion receptors. Therefore, nature-derived polymers can increase the bioactivity of synthetic polymers, when combined with them in nanofibrous scaffolds. Conversely, synthetic polymers can improve the electrospinnability and mechanical properties of the natural polymers. Similarly as synthetic polymers, nature-derived polymers can be more or less degradable in human tissues. Degradable polymers include collagen, gelatin, elastin, keratin, glycosaminoglycans, but also chitosan, gellan gum and PHBV, that is, polymers produced by other than mammalian organisms. Polymers produced by other organisms, such as bacteria, fungi, algae, plants or insects, are usually nondegradable in human tissues, or their degradability is limited due to lack of appropriate enzymes. These polymers include glucans, such as cellulose or dextran, and other polysaccharides and proteins, such as pullulan, alginate, pectin, and silk fibroin. Well-degradable polymers are recommended as direct scaffolds for tissue engineering, while less-degradable polymers are suitable for "intelligent" wound dressing for drug delivery and cell delivery.

Acknowledgements

This review article was supported by the Grant Agency of the Czech Republic (grants No. 17-02448S and 17-00885S).

Author details

Lucie Bacakova[1*], Julia Pajorova[1], Marketa Zikmundova[1], Elena Filova[1], Petr Mikes[2], Vera Jencova[2], Eva Kuzelova Kostakova[2] and Alla Sinica[3,4]

1 Department of Biomaterials and Tissue Engineering, Institute of Physiology of the Czech Academy of Sciences, Prague, Czech Republic

2 Faculty of Sciences, Humanities and Education, Technical University of Liberec, Liberec, Czech Republic

3 BIOCEV, 1st Faculty of Medicine, Charles University, Vestec, Czech Republic

4 Department of Analytical Chemistry, University of Chemistry and Technology Prague, Prague, Czech Republic

*Address all correspondence to: lucie.bacakova@fgu.cas.cz

References

[1] Bacakova L, Bacakova M, Pajorova J, Kudlackova R, Stankova L, Musilkova J, et al. Nanofibrous scaffolds as promising cell carriers for tissue engineering, chapter 3. In: Rahman MM, Asiri AM, editors, ISBN 978-953-51-2529-7, Print ISBN 978-953-51-2528-0, Nanofiber Research - Reaching New Heights. London, United Kingdom: IntechOpen; 2016. pp. 29-54

[2] Kim JI, Pant HR, Sim HJ, Lee KM, Kim CS. Electrospun propolis/polyurethane composite nanofibers for biomedical applications. Materials Science & Engineering. C, Materials for Biological Applications. 2014;**44**:52-57. DOI: 10.1016/j.msec.2014.07.062

[3] Drupitha MP, Bankoti K, Pal P, Das B, Parameswar R, Dhara S, et al. Morphology-induced physico-mechanical and biological characteristics of TPU-PDMS blend scaffolds for skin tissue engineering applications. Journal of Biomedical Materials Research. Part B, Applied Biomaterials. 2019;**107**(5):1634-1644. DOI: 10.1002/jbm.b.34256

[4] Arslan A, Simşek M, Aldemir SD, Kazaroğlu NM, Gümüşderelioğlu M. Honey-based PET or PET/chitosan fibrous wound dressings: Effect of honey on electrospinning process. Journal of Biomaterials Science. Polymer Edition. 2014;**25**(10):999-1012. DOI: 10.1080/09205063.2014.918455

[5] Babaeijandaghi F, Shabani I, Seyedjafari E, Naraghi ZS, Vasei M, Haddadi-Asl V, et al. Accelerated epidermal regeneration and improved dermal reconstruction achieved by polyethersulfone nanofibers. Tissue Engineering. Part A. 2010;**16**(11): 3527-3536. DOI: 10.1089/ten.TEA.2009.0829

[6] Altinbasak I, Jijie R, Barras A, Golba B, Sanyal R, Bouckaert J, et al. Reduced graphene-oxide-embedded polymeric nanofiber mats: An "on-demand" photothermally triggered antibiotic release platform. ACS Applied Materials & Interfaces. 2018;**10**(48):41098-41106. DOI: 10.1021/acsami.8b14784

[7] Zupančič Š, Sinha-Ray S, Sinha-Ray S, Kristl J, Yarin AL. Controlled release of ciprofloxacin from core-shell nanofibers with monolithic or blended core. Molecular Pharmaceutics. 2016;**13**(4):1393-1404. DOI: 10.1021/acs.molpharmaceut.6b00039

[8] Li H, Williams GR, Wu J, Lv Y, Sun X, Wu H, et al. Thermosensitive nanofibers loaded with ciprofloxacin as antibacterial wound dressing materials. International Journal of Pharmaceutics. 2017;**517**(1-2):135-147. DOI: 10.1016/j.ijpharm.2016.12.008

[9] Hejazian LB, Esmaeilzade B, Moghanni Ghoroghi F, Moradi F, Hejazian MB, Aslani A, et al. The role of biodegradable engineered nanofiber scaffolds seeded with hair follicle stem cells for tissue engineering. Iranian Biomedical Journal. 2012;**16**(4):193-201. DOI: 10.6091/ibj.1074.2012

[10] Mikes P, Horakova J, Saman A, Vejsadova L, Topham P, Punyodom W, et al. Comparison and characterization of different polyester nano/micro fibres for use in tissue engineering applications. Journal of Industrial Textiles. 2019. In press. DOI: 10.1177/1528083719848155

[11] Hoveizi E, Nabiuni M, Parivar K, Rajabi-Zeleti S, Tavakol S. Functionalisation and surface modification of electrospun polylactic acid scaffold for tissue engineering. Cell Biology International. 2014;**38**(1):41-49. DOI: 10.1002/cbin.10178

[12] Shtrichman R, Zeevi-Levin N, Zaid R, Barak E, Fishman B,

Ziskind A, et al. The generation of hybrid electrospun nanofiber layer with extracellular matrix derived from human pluripotent stem cells, for regenerative medicine applications. Tissue Engineering. Part A. 2014;**20**(19-20):2756-2767. DOI: 10.1089/ten.TEA.2013.0705

[13] Poormasjedi-Meibod MS, Salimi Elizei S, Leung V, Baradar Jalili R, Ko F, Ghahary A. Kynurenine modulates MMP-1 and type-I collagen expression via aryl hydrocarbon receptor activation in dermal fibroblasts. Journal of Cellular Physiology. 2016;**231**(12):2749-2760. DOI: 10.1002/jcp.25383

[14] Gaaz TS, Sulong AB, Akhtar MN, Kadhum AA, Mohamad AB, Al-Amiery AA. Properties and applications of polyvinyl alcohol, halloysite nanotubes and their nanocomposites. Molecules. 2015;**20**(12):22833-22847. DOI: 10.3390/molecules201219884

[15] Bacakova L, Svorcik V. Cell colonization control by physical and chemical modification of materials. In: Kimura D, editor. Cell Growth Processes: New Research. Hauppauge, New York, USA: Nova Science Publishers, Inc; 2008. pp. 5-56. ISBN 978-1-60456-123-6

[16] Bacakova L, Filova E, Parizek M, Ruml T, Svorcik V. Modulation of cell adhesion, proliferation and differentiation on materials designed for body implants. Biotechnology Advances. 2011;**29**(6):739-767. DOI: 10.1016/j.biotechadv.2011.06.004

[17] Blackstone BN, Hahn JM, McFarland KL, DeBruler DM, Supp DM, Powell HM. Inflammatory response and biomechanical properties of coaxial scaffolds for engineered skin in vitro and post-grafting. Acta Biomaterialia. 2018;**80**:247-257. DOI: 10.1016/j.actbio.2018.09.014

[18] Bačáková L, Novotná K, Sopuch T, Havelka P. Cell interaction with cellulose-based scaffolds for tissue engineering – A review. In: Mondal IH, editor. Cellulose and Cellulose Derivatives: Synthesis, Modification, Nanostructure and Applications. Hauppauge, New York, USA: Nova Science Publishers, Inc.; 2015. pp. 341-375. ISBN 978-1-63483-571-8

[19] Bacakova L, Pajorova J, Bacakova M, Skogberg A, Kallio P, Kolarova K, et al. Versatile application of nanocellulose: From industry to skin tissue engineering and wound healing. Nanomaterials. 2019;**9**(2). pii: E164. DOI: 10.3390/nano9020164

[20] Ohnkawa K. Nanofibers of cellulose and its derivatives fabricated using direct electrospinning. Molecules. 2015;**20**:9139-9154. DOI: 10.3390/molecules20059139

[21] Hu Y, Catchmark JM. Integration of cellulases into bacterial cellulose: Toward bioabsorbable cellulose composites. Journal of Biomedical Materials Research. Part B, Applied Biomaterials. 2011;**97**(1):114-123. DOI: 10.1002/jbm.b.31792

[22] Hu Y, Catchmark JM. In vitro biodegradability and mechanical properties of bioabsorbable bacterial cellulose incorporating cellulases. Acta Biomaterialia. 2011;**7**(7):2835-2845. DOI: 10.1016/j.actbio.2011.03.028

[23] Ko IK, Iwata H. An approach to constructing three-dimensional tissue. Annals of the New York Academy of Sciences. 2001;**944**:443-455. DOI: 10.1111/j.1749-6632.2001.tb03854.x

[24] Ko IK, Kato K, Iwata H. A thin carboxymethyl cellulose culture substrate for the cellulase-induced harvesting of an endothelial cell sheet. Journal of Biomaterials Science. Polymer Edition. 2005;**16**(10):1277-1291. DOI: 10.1163/156856205774269511

[25] Entcheva E, Bien H, Yin L, Chung CY, Farrell M, Kostov Y. Functional cardiac cell constructs on cellulose-based scaffolding. Biomaterials. 2004;**25**(26):5753-5762. DOI: 10.1016/j.biomaterials.2004.01.024

[26] Torres-Rendon JG, Köpf M, Gehlen D, Blaeser A, Fischer H, De Laporte L, et al. Cellulose nanofibril hydrogel tubes as sacrificial templates for freestanding tubular cell constructs. Biomacromolecules. 2016;**17**(3):905-913. DOI: 10.1021/acs.biomac.5b01593

[27] Yadav V, Paniliatis BJ, Shi H, Lee K, Cebe P, Kaplan DL. Novel in vivo-degradable cellulose-chitin copolymer from metabolically engineered Gluconacetobacter xylinus. Applied and Environmental Microbiology. 2010;**76**(18):6257-6265. DOI: 10.1128/AEM.00698-10

[28] Yadav V, Sun L, Panilaitis B, Kaplan DL. In vitro chondrogenesis with lysozyme susceptible bacterial cellulose as a scaffold. Journal of Tissue Engineering and Regenerative Medicine. 2015;**9**(12):E276-E288. DOI: 10.1002/term.1644

[29] Safaee-Ardakani MR, Hatamian-Zarmi A, Sadat SM, Mokhtari-Hosseini ZB, Ebrahimi-Hosseinzadeh B, Rashidiani J, et al. Electrospun schizophyllan/polyvinyl alcohol blend nanofibrous scaffold as potential wound healing. International Journal of Biological Macromolecules. 2019;**127**:27-38. DOI: 10.1016/j.ijbiomac.2018.12.256

[30] Unnithan AR, Sasikala AR, Murugesan P, Gurusamy M, Wu D, Park CH, et al. Electrospun polyurethane-dextran nanofiber mats loaded with estradiol for post-menopausal wound dressing. International Journal of Biological Macromolecules. 2015;**77**:1-8. DOI: 10.1016/j.ijbiomac.2015.02.044

[31] Pan JF, Liu NH, Sun H, Xu F. Preparation and characterization of electrospun PLCL/Poloxamer nanofibers and dextran/gelatin hydrogels for skin tissue engineering. PLoS One. 2014;**9**(11):e112885. DOI: 10.1371/journal.pone.0112885

[32] Maia J, Evangelista MB, Gil H, Ferreira L. 2. Dextran-based materials for biomedical applications. In: Gil MH, editor. Carbohydrates Applications in Medicine. Kerala, India: Research Signpost; 2014. pp. 31-53. ISBN: 978-81-308-0523-8

[33] Bačáková L, Novotná K, Pařízek M. Polysaccharides as cell carriers for tissue engineering: The use of cellulose in vascular wall reconstruction. Physiological Research. 2014;**63**(Suppl 1):S29-S47

[34] Waghmare VS, Wadke PR, Dyawanapelly S, Deshpande A, Jain R, Dandekar P. Starch based nanofibrous scaffolds for wound healing applications. Bioact Mater. 2017;**3**(3):255-266. DOI: 10.1016/j.bioactmat.2017.11.006

[35] Byman E, Schultz N, Netherlands Brain Bank, Fex M, Wennström M. Brain alpha-amylase: A novel energy regulator important in Alzheimer disease? Brain Pathology. 2018;**28**(6):920-932. DOI: 10.1111/bpa.12597

[36] Abed A, Assoul N, Ba M, Derkaoui SM, Portes P, Louedec L, et al. Influence of polysaccharide composition on the biocompatibility of pullulan/dextran-based hydrogels. Journal of Biomedical Materials Research. Part A. 2011;**96**(3):535-542. DOI: 10.1002/jbm.a.33007

[37] Xu F, Weng B, Gilkerson R, Materon LA, Lozano K. Development of tannic acid/chitosan/pullulan composite nanofibers from aqueous solution for potential applications as wound dressing. Carbohydrate Polymers.

2015;**115**:16-24. DOI: 10.1016/j.carbpol.2014.08.081

[38] Shi L, Le Visage C, Chew SY. Long-term stabilization of polysaccharide electrospun fibres by in situ cross-linking. Journal of Biomaterials Science. Polymer Edition. 2011;**22**(11):1459-1472. DOI: 10.1163/092050610X512108

[39] Krishnan R, Rajeswari R, Venugopal J, Sundarrajan S, Sridhar R, Shayanti M, et al. Polysaccharide nanofibrous scaffolds as a model for in vitro skin tissue regeneration. Journal of Materials Science. Materials in Medicine. 2012;**23**(6):1511-1519. DOI: 10.1007/s10856-012-4630-6

[40] Liu X, Chang M, He B, Meng L, Wang X, Sun R, et al. A one-pot strategy for preparation of high-strength carboxymethyl xylan-g-poly(acrylic acid) hydrogels with shape memory property. Journal of Colloid and Interface Science. 2019;**538**:507-518. DOI: 10.1016/j.jcis.2018.12.023

[41] Sobhanian P, Khorram M, Hashemi SS, Mohammadi A. Development of nanofibrous collagen-grafted poly (vinyl alcohol)/gelatin/alginate scaffolds as potential skin substitute. International Journal of Biological Macromolecules. 2019;**130**:977-987. DOI: 10.1016/j.ijbiomac.2019.03.045

[42] Coşkun G, Karaca E, Ozyurtlu M, Ozbek S, Yermezler A, Cavuşoğlu I. Histological evaluation of wound healing performance of electrospun poly(vinyl alcohol)/sodium alginate as wound dressing in vivo. Bio-medical Materials and Engineering. 2014;**24**(2):1527-1536. DOI: 10.3233/BME-130956

[43] Jeong SI, Krebs MD, Bonino CA, Khan SA, Alsberg E. Electrospun alginate nanofibers with controlled cell adhesion for tissue engineering. Macromolecular Bioscience. 2010;**10**(8):934-943. DOI: 10.1002/mabi.201000046

[44] Tort S, Acartürk F, Beşikci A. Evaluation of three-layered doxycycline-collagen loaded nanofiber wound dressing. International Journal of Pharmaceutics. 2017;**529**(1-2):642-653. DOI: 10.1016/j.ijpharm.2017.07.027

[45] Byeon SY, Cho MK, Shim KH, Kim HJ, Song HG, Shin HS. Development of a Spirulina extract/alginate-imbedded PCL nanofibrous cosmetic patch. Journal of Microbiology and Biotechnology. 2017;**27**(9):1657-1663. DOI: 10.4014/jmb.1701.01025

[46] Hoseinpour Najar M, Minaiyan M. Taheri A. Preparation and in vivo evaluation of a novel gel-based wound dressing using arginine-alginate surface-modified chitosan nanofibers. Journal of Biomaterials Applications. 2018;**32**(6):689-701. DOI: 10.1177/0885328217739562

[47] Hu WP, Zhang B, Zhang J, Luo WL, Guo Y, Chen SJ, et al. Ag/alginate nanofiber membrane for flexible electronic skin. Nanotechnology. 2017;**28**(44):445502. DOI: 10.1088/1361-6528/aa8746

[48] Guarino V, Caputo T, Altobelli R, Ambrosio L. Degradation properties and metabolic activity of alginate and chitosan polyelectrolytes for drug delivery and tissue engineering applications. AIMS Materials Science. 2015;**2**(4):497-502. DOI: 10.3934/matersci.2015.4.497

[49] Voragen AGJ, Coenen G-J, Verhoef RP, Schols HA. Pectin, a versatile polysaccharide present in plant cell walls. Structural Chemistry. 2009;**20**:263-275. DOI: 10.1007/s11224-009-9442-z

[50] Archana D, Dutta J, Dutta PK. Evaluation of chitosan nano dressing

for wound healing: Characterization, in vitro and in vivo studies. International Journal of Biological Macromolecules. 2013;**57**:193-203. DOI: 10.1016/j.ijbiomac.2013.03.002

[51] Tayi L, Maku RV, Patel HK, Sonti RV. Identification of pectin degrading enzymes secreted by Xanthomonas oryzae pv. oryzae and determination of their role in virulence on rice. PLoS One. 2016;**11**(12):e0166396. DOI: 10.1371/journal.pone.0166396

[52] Martos MA, Zubreski ER, Garro OA, Hours RA. Production of pectinolytic enzymes by the yeast Wickerhanomyces anomalus isolated from citrus fruits peels. Biotechnology Research International. 2013;**2013**:435154. DOI: 10.1155/2013/435154

[53] Huber DJ, Karakurt Y, Jeong J, et al. Pectin degradation in ripening and wounded fruits. The Revista Brasileira de Fisiologia Vegetal (Brazilian Journal of Plant Physiology). 2001;**13**(2):224-241. DOI: 10.1590/S0103-31312001000200009

[54] Lin HY, Chen HH, Chang SH, Ni TS. Pectin-chitosan-PVA nanofibrous scaffold made by electrospinning and its potential use as a skin tissue scaffold. Journal of Biomaterials Science. Polymer Edition. 2013;**24**(4):470-484. DOI: 10.1080/09205063.2012.693047

[55] Lin HY, Chen SH, Chang SH, Huang ST. Tri-layered chitosan scaffold as a potential skin substitute. Journal of Biomaterials Science. Polymer Edition. 2015;**26**(13):855-867. DOI: 10.1080/09205063.2015.1061350

[56] Chen S, Cui S, Zhang H, Pei X, Hu J, Zhou Y, et al. Cross-linked pectin nanofibers with enhanced cell adhesion. Biomacromolecules. 2018;**19**(2):490-498. DOI: 10.1021/acs.biomac.7b01605

[57] Saruchi, Kaith BS, Jindal R, Kumar V. Biodegradation of gum tragacanth acrylic acid based hydrogel and its impact on soil fertility. Polymer Degradation and Stability. 2015;**115**:24-31. DOI: 10.1016/j.polymdegradstab.2015.02.009

[58] Hindi SSZ, Albureikan MO, Al-Ghamdi AA, Alhummiany H, Ansari MS. Synthesis, characterization and biodegradation of gum arabic-based bioplastic membranes. Nanoscience and Nanotechnology Research. 2017;**4**(2):32-42. DOI: 10.12691/nnr-4-2-1

[59] Zarekhalili Z, Bahrami SH, Ranjbar-Mohammadi M, Milan PB. Fabrication and characterization of PVA/gum tragacanth/PCL hybrid nanofibrous scaffolds for skin substitutes. International Journal of Biological Macromolecules. 2017;**94**(Pt A):679-690. DOI: 10.1016/j.ijbiomac.2016.10.042

[60] Ranjbar-Mohammadi M, Rabbani S, Bahrami SH, Joghataei MT, Moayer F. Antibacterial performance and in vivo diabetic wound healing of curcumin loaded gum tragacanth/poly(ε-caprolactone) electrospun nanofibers. Materials Science & Engineering. C, Materials for Biological Applications. 2016;**69**:1183-1191. DOI: 10.1016/j.msec.2016.08.032

[61] Pedram Rad Z, Mokhtari J, Abbasi M. Calendula officinalis extract/PCL/Zein/gum arabic nanofibrous bio-composite scaffolds via suspension, two-nozzle and multilayer electrospinning for skin tissue engineering. International Journal of Biological Macromolecules. 2019;**135**:530-543. DOI: 10.1016/j.ijbiomac.2019.05.204

[62] Srivastava CM, Purwar R. Fabrication of robust Antheraea assama fibroin nanofibrous mat using ionic liquid for skin tissue engineering.

Materials Science & Engineering. C, Materials for Biological Applications. 2016;**68**:276-290. DOI: 10.1016/j.msec.2016.05.020

[63] Srivastava CM, Purwar R, Gupta AP. Enhanced potential of biomimetic, silver nanoparticles functionalized Antheraea mylitta (tasar) silk fibroin nanofibrous mats for skin tissue engineering. International Journal of Biological Macromolecules. 2019;**130**:437-453. DOI: 10.1016/j.ijbiomac.2018.12.255

[64] Chouhan D, Chakraborty B, Nandi SK, Mandal BB. Role of non-mulberry silk fibroin in deposition and regulation of extracellular matrix towards accelerated wound healing. Acta Biomaterialia. 2017;**48**:157-174. DOI: 10.1016/j.actbio.2016.10.019

[65] Cao Y, Wang B. Biodegradation of silk biomaterials. International Journal of Molecular Sciences. 2009;**10**(4):1514-1524. DOI: 10.3390/ijms10041514

[66] Wang Y, Rudym DD, Walsh A, Abrahamsen L, Kim HJ, Kim HS, et al. In vivo degradation of three-dimensional silk fibroin scaffolds. Biomaterials. 2008;**29**(24-25): 3415-3428. DOI: 10.1016/j.biomaterials.2008.05.002

[67] Lee JM, Chae T, Sheikh FA, Ju HW, Moon BM, Park HJ, et al. Three dimensional poly(ε-caprolactone) and silk fibroin nanocomposite fibrous matrix for artificial dermis. Materials Science & Engineering. C, Materials for Biological Applications. 2016;**68**:758-767. DOI: 10.1016/j.msec.2016.06.019

[68] Sridhar S, Venugopal JR, Ramakrishna S. Improved regeneration potential of fibroblasts using ascorbic acid-blended nanofibrous scaffolds. Journal of Biomedical Materials Research. Part A. 2015;**103**(11):3431-3440. DOI: 10.1002/jbm.a.35486

[69] Zhou Y, Yang H, Liu X, Mao J, Gu S, Xu W. Electrospinning of carboxyethyl chitosan/poly(vinyl alcohol)/silk fibroin nanoparticles for wound dressings. International Journal of Biological Macromolecules. 2013;**53**:88-92. DOI: 10.1016/j.ijbiomac.2012.11.013

[70] Yoo CR, Yeo IS, Park KE, Park JH, Lee SJ, Park WH, et al. Effect of chitin/silk fibroin nanofibrous bicomponent structures on interaction with human epidermal keratinocytes. International Journal of Biological Macromolecules. 2008;**42**(4):324-334. DOI: 10.1016/j.ijbiomac.2007.12.004

[71] Shefa AA, Amirian J, Kang HJ, Bae SH, Jung HI, Choi HJ, et al. In vitro and in vivo evaluation of effectiveness of a novel TEMPO-oxidized cellulose nanofiber-silk fibroin scaffold in wound healing. Carbohydrate Polymers. 2017;**177**:284-296. DOI: 10.1016/j.carbpol.2017.08.130

[72] Xin S, Li X, Wang Q, Huang R, Xu X, Lei Z, et al. Novel layer-by-layer structured nanofibrous mats coated by protein films for dermal regeneration. Journal of Biomedical Nanotechnology. 2014;**10**(5):803-810. DOI: 10.1166/jbn.2014.1748

[73] Yeo IS, Oh JE, Jeong L, Lee TS, Lee SJ, Park WH, et al. Collagen-based biomimetic nanofibrous scaffolds: Preparation and characterization of collagen/silk fibroin bicomponent nanofibrous structures. Biomacromolecules. 2008;**9**(4): 1106-1116. DOI: 10.1021/bm700875a

[74] Shan YH, Peng LH, Liu X, Chen X, Xiong J, Gao JQ. Silk fibroin/gelatin electrospun nanofibrous dressing functionalized with astragaloside IV induces healing and anti-scar effects on burn wound. International Journal of Pharmaceutics. 2015;**479**(2):291-301. DOI: 10.1016/j.ijpharm.2014.12.067

[75] Li L, Qian Y, Jiang C, Lv Y, Liu W, Zhong L, et al. The use of hyaluronan

to regulate protein adsorption and cell infiltration in nanofibrous scaffolds. Biomaterials. 2012;**33**(12):3428-3445. DOI: 10.1016/j.biomaterials.2012.01.038

[76] Sheng X, Fan L, He C, Zhang K, Mo X, Wang H. Vitamin E-loaded silk fibroin nanofibrous mats fabricated by green process for skin care application. International Journal of Biological Macromolecules. 2013;**56**:49-56. DOI: 10.1016/j.ijbiomac.2013.01.029

[77] Fan L, Cai Z, Zhang K, Han F, Li J, He C, et al. Green electrospun pantothenic acid/silk fibroin composite nanofibers: Fabrication, characterization and biological activity. Colloids and Surfaces. B, Biointerfaces. 2014;**117**:14-20. DOI: 10.1016/j.colsurfb.2013.12.030

[78] Lin S, Chen M, Jiang H, Fan L, Sun B, Yu F, et al. Green electrospun grape seed extract-loaded silk fibroin nanofibrous mats with excellent cytocompatibility and antioxidant effect. Colloids and Surfaces. B, Biointerfaces. 2016;**139**:156-163. DOI: 10.1016/j.colsurfb.2015.12.001

[79] Kandhasamy S, Arthi N, Arun RP, Verma RS. Synthesis and fabrication of novel quinone-based chromenopyrazole antioxidant-laden silk fibroin nanofibers scaffold for tissue engineering applications. Materials Science & Engineering. C, Materials for Biological Applications. 2019;**102**:773-787. DOI: 10.1016/j.msec.2019.04.076

[80] Suganya S, Venugopal J, Ramakrishna S, Lakshmi BS, Dev VR. Naturally derived biofunctional nanofibrous scaffold for skin tissue regeneration. International Journal of Biological Macromolecules. 2014;**68**:135-143. DOI: 10.1016/j.ijbiomac.2014.04.031

[81] Lee OJ, Ju HW, Kim JH, Lee JM, Ki CS, Kim JH, et al. Development of artificial dermis using 3D electrospun silk fibroin nanofiber matrix. Journal of Biomedical Nanotechnology. 2014;**10**(7):1294-1303

[82] Park YR, Ju HW, Lee JM, Kim DK, Lee OJ, Moon BM, et al. Three-dimensional electrospun silk-fibroin nanofiber for skin tissue engineering. International Journal of Biological Macromolecules. 2016;**93**(Pt B):1567-1574. DOI: 10.1016/j.ijbiomac.2016.07.047

[83] Gholipourmalekabadi M, Samadikuchaksaraei A, Seifalian AM, Urbanska AM, Ghanbarian H, Hardy JG, et al. Silk fibroin/amniotic membrane 3D bi-layered artificial skin. Biomedical Materials. 2018;**13**(3):035003. DOI: 10.1088/1748-605X/aa999b

[84] Xie SY, Peng LH, Shan YH, Niu J, Xiong J, Gao JQ. Adult stem cells seeded on electrospinning silk fibroin nanofiberous scaffold enhance wound repair and regeneration. Journal of Nanoscience and Nanotechnology. 2016;**16**(6):5498-5505. DOI: 10.1166/jnn.2016.11730

[85] Sapru S, Das S, Mandal M, Ghosh AK, Kundu SC. Prospects of nonmulberry silk protein sericin-based nanofibrous matrices for wound healing - In vitro and in vivo investigations. Acta Biomaterialia. 2018;**78**:137-150. DOI: 10.1016/j.actbio.2018.07.047

[86] Bhowmick S, Scharnweber D, Koul V. Co-cultivation of keratinocyte-human mesenchymal stem cell (hMSC) on sericin loaded electrospun nanofibrous composite scaffold (cationic gelatin/hyaluronan/chondroitin sulfate) stimulates epithelial differentiation in hMSCs: In vitro study. Biomaterials. 2016;**88**:83-96. DOI: 10.1016/j.biomaterials.2016.02.034

[87] Gilotra S, Chouhan D, Bhardwaj N, Nandi SK, Mandal BB. Potential of silk sericin based nanofibrous mats

for wound dressing applications. Materials Science & Engineering. C, Materials for Biological Applications. 2018;**90**:420-432. DOI: 10.1016/j.msec.2018.04.077

[88] Zhao R, Li X, Sun B, Zhang Y, Zhang D, Tang Z, et al. Electrospun chitosan/sericin composite nanofibers with antibacterial property as potential wound dressings. International Journal of Biological Macromolecules. 2014;**68**:92-97. DOI: 10.1016/j.ijbiomac.2014.04.029

[89] Law JX, Liau LL, Saim A, Yang Y, Idrus R. Electrospun collagen nanofibers and their applications in skin tissue engineering. Tissue Eng Regen Med. 2017;**14**(6):699-718. DOI: 10.1007/s13770-017-0075-9

[90] Wang Y, Zhang CL, Zhang Q, Li P. Composite electrospun nanomembranes of fish scale collagen peptides/chito-oligosaccharides: Antibacterial properties and potential for wound dressing. International Journal of Nanomedicine. 2011;**6**: 667-676. DOI: 10.2147/IJN.S17547

[91] Zhou T, Wang N, Xue Y, Ding T, Liu X, Mo X, et al. Electrospun tilapia collagen nanofibers accelerating wound healing via inducing keratinocytes proliferation and differentiation. Colloids and Surfaces. B, Biointerfaces. 2016;**143**:415-422. DOI: 10.1016/j.colsurfb.2016.03.052

[92] Le Corre-Bordes D, Hofman K, Hall B. Guide to electrospinning denatured whole chain collagen from hoki fish using benign solvents. International Journal of Biological Macromolecules. 2018;**112**:1289-1299. DOI: 10.1016/j.ijbiomac.2018.02.088

[93] Zhou T, Sui B, Mo X, Sun J. Multifunctional and biomimetic fish collagen/bioactive glass nanofibers: Fabrication, antibacterial activity and inducing skin regeneration in vitro and in vivo. International Journal of Nanomedicine. 2017;**12**:3495-3507. DOI: 10.2147/IJN.S132459

[94] Chen J, Gao K, Liu S, Wang S, Elango J, Bao B, et al. Fish collagen surgical compress repairing characteristics on wound healing process In vivo. Marine Drugs. 2019;**17**. pii: E33. DOI: 10.3390/md17010033

[95] Cumming MH, Leonard AR, LeCorre-Bordes DS, Hofman K. Intra-fibrillar citric acid crosslinking of marine collagen electrospun nanofibres. International Journal of Biological Macromolecules. 2018;**114**:874-881. DOI: 10.1016/j.ijbiomac.2018.03.180

[96] Ramanathan G, Muthukumar T, Sivagnanam T. In vivo efficiency of the collagen coated nanofibrous scaffold and their effect on growth factors and pro-inflammatory cytokines in wound healing. European Journal of Pharmacology. 2017;**814**:45-55. DOI: 10.1016/j.ejphar.2017.08.003

[97] Dhand C, Balakrishnan Y, Ong ST, Dwivedi N, Venugopal JR, Harini S, et al. Antimicrobial quaternary ammonium organosilane cross-linked nanofibrous collagen scaffolds for tissue engineering. International Journal of Nanomedicine. 2018;**13**:4473-4492. DOI: 10.2147/IJN.S159770

[98] Ravichandran R, Venugopal JR, Sundarrajan S, Mukherjee S, Sridhar R, Ramakrishna S. Composite poly-L-lactic acid/poly-(α,β)-DL-aspartic acid/collagen nanofibrous scaffolds for dermal tissue regeneration. Materials Science & Engineering. C, Materials for Biological Applications. 2012;**32**(6):1443-1451. DOI: 10.1016/j.msec.2012.04.024

[99] Alamein MA, Stephens S, Liu Q, Skabo S, Warnke PH. Mass production of nanofibrous extracellular matrix with controlled 3D morphology for

large-scale soft tissue regeneration. Tissue Engineering. Part C, Methods. 2013;**19**(6):458-472. DOI: 10.1089/ten.TEC.2012.0417

[100] Peh P, Lim NS, Blocki A, Chee SM, Park HC, Liao S, et al. Simultaneous delivery of highly diverse bioactive compounds from blend electrospun fibers for skin wound healing. Bioconjugate Chemistry. 2015;**26**(7):1348-1358. DOI: 10.1021/acs.bioconjchem.5b00123

[101] Gümüşderelioğlu M, Dalkıranoğlu S, Aydın RS, Cakmak S. A novel dermal substitute based on biofunctionalized electrospun PCL nanofibrous matrix. Journal of Biomedical Materials Research. Part A. 2011;**98**(3):461-472. DOI: 10.1002/jbm.a.33143

[102] Albright V, Xu M, Palanisamy A, Cheng J, Stack M, Zhang B, et al. Micelle-coated, hierarchically structured nanofibers with dual-release capability for accelerated wound healing and infection control. Advanced Healthcare Materials. 2018;**7**(11):e1800132. DOI: 10.1002/adhm.201800132

[103] Abdul Khodir WKW, Abdul Razak AH, Ng MH, Guarino V, Susanti D. Encapsulation and characterization of gentamicin sulfate in the collagen added electrospun nanofibers for skin regeneration. J Funct Biomater. 2018;**9**(2):E36. DOI: 10.3390/jfb9020036

[104] Wang C, Wang Q, Gao W, Zhang Z, Lou Y, Jin H, et al. Highly efficient local delivery of endothelial progenitor cells significantly potentiates angiogenesis and full-thickness wound healing. Acta Biomaterialia. 2018;**69**:156-169. DOI: 10.1016/j.actbio.2018.01.019

[105] Ghosal K, Manakhov A, Zajíčková L, Thomas S. Structural and surface compatibility study of modified electrospun poly(ε-caprolactone) (PCL) composites for skin tissue engineering. AAPS PharmSciTech. 2017;**18**(1):72-81. DOI: 10.1208/s12249-016-0500-8

[106] Xie X, Li D, Su C, Cong W, Mo X, Hou G, et al. Functionalized biomimetic composite nanfibrous scaffolds with antibacterial and hemostatic efficacy for facilitating wound healing. Journal of Biomedical Nanotechnology. 2019;**15**(6):1267-1279. DOI: 10.1166/jbn.2019.2756

[107] Yao CH, Chen KY, Chen YS, Li SJ, Huang CH. Lithospermi radix extract-containing bilayer nanofiber scaffold for promoting wound healing in a rat model. Materials Science & Engineering. C, Materials for Biological Applications. 2019;**96**:850-858. DOI: 10.1016/j.msec.2018.11.053

[108] Aldana AA, Abraham GA. Current advances in electrospun gelatin-based scaffolds for tissue engineering applications. International Journal of Pharmaceutics. 2017;**523**(2):441-453. DOI: 10.1016/j.ijpharm.2016.09.044

[109] Yao CH, Lee CY, Huang CH, Chen YS, Chen KY. Novel bilayer wound dressing based on electrospun gelatin/keratin nanofibrous mats for skin wound repair. Materials Science & Engineering. C, Materials for Biological Applications. 2017;**79**:533-540. DOI: 10.1016/j.msec.2017.05.076

[110] Adeli-Sardou M, Yaghoobi MM, Torkzadeh-Mahani M, Dodel M. Controlled release of lawsone from polycaprolactone/gelatin electrospun nano fibers for skin tissue regeneration. International Journal of Biological Macromolecules. 2019;**124**:478-491. DOI: 10.1016/j.ijbiomac.2018.11.237

[111] Baghersad S, Hajir Bahrami S, Mohammadi MR, Mojtahedi MRM, Milan PB. Development of biodegradable electrospun gelatin/aloe-vera/poly (ε-caprolactone) hybrid nanofibrous

scaffold for application as skin substitutes. Materials Science & Engineering. C, Materials for Biological Applications. 2018;**93**:367-379. DOI: 10.1016/j.msec.2018.08.020

[112] Dias JR, Baptista-Silva S, Sousa A, Oliveira AL, Bártolo PJ, Granja PL. Biomechanical performance of hybrid electrospun structures for skin regeneration. Materials Science & Engineering. C, Materials for Biological Applications. 2018;**93**:816-827. DOI: 10.1016/j.msec.2018.08.050

[113] Chandrasekaran AR, Venugopal J, SundarrajanS, RamakrishnaS. Fabrication of a nanofibrous scaffold with improved bioactivity for culture of human dermal fibroblasts for skin regeneration. Biomedical Materials. 2011;**6**(1):015001. DOI: 10.1088/1748-6041/6/1/015001

[114] Jin G, Prabhakaran MP, Kai D, Ramakrishna S. Controlled release of multiple epidermal induction factors through core-shell nanofibers for skin regeneration. European Journal of Pharmaceutics and Biopharmaceutics. 2013;**85**(3 Pt A):689-698. DOI: 10.1016/j.ejpb.2013.06.002

[115] Mao W, Kang MK, Shin JU, Son YJ, Kim HS, Yoo HS. Coaxial hydro-nanofibrils for self-assembly of cell sheets producing skin bilayers. ACS Applied Materials & Interfaces. 2018;**10**(50):43503-43511. DOI: 10.1021/acsami.8b17740

[116] Kim JW, Kim MJ, Ki CS, Kim HJ, Park YH. Fabrication of bi-layer scaffold of keratin nanofiber and gelatin-methacrylate hydrogel: Implications for skin graft. International Journal of Biological Macromolecules. 2017;**105**(Pt 1):541-548. DOI: 10.1016/j.ijbiomac.2017.07.067

[117] Zhao X, Sun X, Yildirimer L, Lang Q, Lin ZYW, Zheng R, et al. Cell infiltrative hydrogel fibrous scaffolds for accelerated wound healing. Acta Biomaterialia. 2017;**49**:66-77. DOI: 10.1016/j.actbio.2016.11.017

[118] DeFrates KG, Moore R, Borgesi J, Lin G, Mulderig T, Beachley V, et al. Protein-based fiber materials in medicine: A review. Nanomaterials. 2018;**8**(7). pii: E457. DOI: 10.3390/nano8070457

[119] Khalili S, Khorasani SN, Razavi SM, Hashemibeni B, Tamayol A. Nanofibrous scaffolds with biomimetic composition for skin regeneration. Applied Biochemistry and Biotechnology. 2019;**187**(4):1193-1203. DOI: 10.1007/s12010-018-2871-7

[120] Riedel T, Brynda E, Dyr JE, Houska M. Controlled preparation of thin fibrin films immobilized at solid surfaces. Journal of Biomedical Materials Research. Part A. 2009;**88**(2):437-447. DOI: 10.1002/jbm.a.31755

[121] Law JX, Musa F, Ruszymah BH, El Haj AJ, Yang Y. A comparative study of skin cell activities in collagen and fibrin constructs. Medical Engineering & Physics. 2016;**38**(9):854-861. DOI: 10.1016/j.medengphy.2016.05.017

[122] Pajorova J, Bacakova M, Musilkova J, Broz A, Hadraba D, Lopot F, et al. Morphology of a fibrin nanocoating influences dermal fibroblast behavior. International Journal of Nanomedicine. 2018;**13**:3367-3380. DOI: 10.2147/IJN.S162644

[123] Bacakova M, Pajorova J, Broz A, Hadraba D, Lopot F, Zavadakova A, et al. A two-layer skin construct consisting of a collagen hydrogel reinforced by a fibrin-coated polylactide nanofibrous membrane. International Journal of Nanomedicine. 2019;**14**:5033-5050. DOI: 10.2147/IJN.S200782

[124] Mirzaei-Parsa MJ, Ghanbari H, Alipoor B, Tavakoli A, Najafabadi MRH, Faridi-Majidi R. Nanofiber-acellular

dermal matrix as a bilayer scaffold containing mesenchymal stem cell for healing of full-thickness skin wounds. Cell and Tissue Research. 2019;**375**(3):709-721. DOI: 10.1007/s00441-018-2927-6

[125] Wang J, Hao S, Luo T, Zhou T, Yang X, Wang B. Keratose/poly (vinyl alcohol) blended nanofibers: Fabrication and biocompatibility assessment. Materials Science & Engineering. C, Materials for Biological Applications. 2017;**72**:212-219. DOI: 10.1016/j.msec.2016.11.071

[126] Giuri D, Barbalinardo M, Sotgiu G, Zamboni R, Nocchetti M, Donnadio A, et al. Nano-hybrid electrospun non-woven mats made of wool keratin and hydrotalcites as potential bio-active wound dressings. Nanoscale. 2019;**11**(13):6422-6430. DOI: 10.1039/c8nr10114k

[127] Cruz-Maya I, Guarino V, Almaguer-Flores A, Alvarez-Perez MA, Varesano A, Vineis C. Highly polydisperse keratin rich nanofibers: Scaffold design and in vitro characterization. Journal of Biomedical Materials Research. Part A. 2019;**107**(8):1803-1813. DOI: 10.1002/jbm.a.36699

[128] Zahedi E, Esmaeili A, Eslahi N, Shokrgozar MA, Simchi A. Fabrication and characterization of core-shell electrospun fibrous mats containing medicinal herbs for wound healing and skin tissue engineering. Marine Drugs. 2019;**17**(1). pii: E27. DOI: 10.3390/md17010027

[129] Yan S, Zhang Q, Wang J, Liu Y, Lu S, Li M, et al. Silk fibroin/chondroitin sulfate/hyaluronic acid ternary scaffolds for dermal tissue reconstruction. Acta Biomaterialia. 2013;**9**(6):6771-6782. DOI: 10.1016/j.actbio.2013.02.016

[130] Li J, He A, Han CC, Fang D, Hsiao BS, Chu B. Electrospinning of hyaluronic acid (HA) and HA/gelatin blends. Macromolecular Rapid Communications. 2006;**27**:114-120. DOI: 10.1002/marc.200500726

[131] Séon-Lutz M, Couffin A-C, Vignoud S, Schlatter G, Hébraud A. Electrospinning in water and in situ crosslinking of hyaluronic acid / cyclodextrin nanofibers: Towards wound dressing with controlled drug release. Carbohydrate Polymers. 2019;**207**:276-287. DOI: 10.1016/j.carbpol.2018.11.085

[132] Um IC, Fang D, Hsiao BS, Okamoto A, Chu B. Electro-spinning and electro-blowing of hyaluronic acid. Biomacromolecules. 2004;**5**(4):1428-1436. DOI: 10.1021/bm034539b

[133] Pokorny M, Rassushin V, Wolfova L, Velebny V. Increased production of nanofibrous materials by electroblowing from blends of hyaluronic acid and polyethylene oxide. Polymer Engineering and Science. 2016;**56**:932-938. DOI: 10.1002/pen.24322

[134] Qian Y, Li L, Jiang C, Xu W, Lv Y, Zhong L, et al. The effect of hyaluronan on the motility of skin dermal fibroblasts in nanofibrous scaffolds. International Journal of Biological Macromolecules. 2015;**79**:133-143. DOI: 10.1016/j.ijbiomac.2015.04.059

[135] Stodolak-Zych E, Rozmus K, Dzierzkowska E, Zych Ł, Rapacz-Kmita A, Gargas M, et al. The membrane with polylactide and hyaluronic fibers for skin substitute. Acta of Bioengineering and Biomechanics. 2018;**20**(4):91-99. DOI: 10.5277/ABB-01199-2018-02

[136] Kurpinski KT, Stephenson JT, Janairo RR, Lee H, Li S. The effect of fiber alignment and heparin coating on cell infiltration into nanofibrous PLLA scaffolds. Biomaterials. 2010;**31**(13):3536-3542. DOI: 10.1016/j.biomaterials.2010.01.062

[137] Yergoz F, Hastar N, Cimenci CE, Ozkan AD, Tekinay T, Guler MO, et al. Heparin mimetic peptide nanofiber gel promotes regeneration of full thickness burn injury. Biomaterials. 2017;**134**:117-127. DOI: 10.1016/j.biomaterials.2017.04.040

[138] Sadeghi A, Zandi M, Pezeshki-Modaress M, Rajabi S. Tough, hybrid chondroitin sulfate nanofibers as a promising scaffold for skin tissue engineering. International Journal of Biological Macromolecules. 2019;**132**: 63-75. DOI: 10.1016/j.ijbiomac.2019.03.208. 27

[139] Pezeshki-Modaress M, Mirzadeh H, Zandi M, Rajabi-Zeleti S, Sodeifi N, Aghdami N, et al. Gelatin/chondroitin sulfate nanofibrous scaffolds for stimulation of wound healing: In-vitro and in-vivo study. Journal of Biomedical Materials Research. Part A. 2017;**105**(7):2020-2034. DOI: 10.1002/jbm.a.35890

[140] Azuma K, Ifuku S, Osaki T, Okamoto Y, Minami S. Preparation and biomedical applications of chitin and chitosan nanofibers. Journal of Biomedical Nanotechnology. 2014;**10**(10):2891-2920. DOI: 10.1166/jbn.2014.1882

[141] Kai D, Liow SS, Loh XJ. Biodegradable polymers for electrospinning: towards biomedical applications. Materials Science and Engineering C: Materials for Biological Applications. 2014;**45**:659-670. DOI: 10.1016/j.msec.2014.04.051

[142] Pal P, Dadhich P, Srivas PK, Das B, Maulik D, Dhara S. Bilayered nanofibrous 3D hierarchy as skin rudiment by emulsion electrospinning for burn wound management. Biomaterials Science. 2017;**5**(9): 1786-1799. DOI: 10.1039/c7bm00174f

[143] Gomes S, Rodrigues G, Martins G, Henriques C, Silva JC. Evaluation of nanofibrous scaffolds obtained from blends of chitosan, gelatin and polycaprolactone for skin tissue engineering. International Journal of Biological Macromolecules. 2017;**102**:1174-1185. DOI: 10.1016/j.ijbiomac.2017.05.004

[144] Mahdavi M, Mahmoudi N, Rezaie Anaran F, Simchi A. Electrospinning of nanodiamond-modified polysaccharide nanofibers with physico-mechanical properties close to natural skins. Marine Drugs. 2016;**14**(7). pii: E128. DOI: 10.3390/md14070128

[145] Veleirinho B, Coelho DS, Dias PF, Maraschin M, Ribeiro-do-Valle RM, Lopes-da-Silva JA. Nanofibrous poly (3-hydroxybutyrate-co-3-hydroxyvalerate)/chitosan scaffolds for skin regeneration. International Journal of Biological Macromolecules. 2012;**51**(4):343-350. DOI: 10.1016/j.ijbiomac.2012.05.023

[146] Shalumon KT, Sathish D, Nair SV, Chennazhi KP, Tamura H, Jayakumar R. Fabrication of aligned poly(lactic acid)-chitosan nanofibers by novel parallel blade collector method for skin tissue engineering. Journal of Biomedical Nanotechnology. 2012;**8**(3):405-416. DOI: 10.1166/jbn.2012.1395

[147] Mo Y, Guo R, Zhang Y, Xue W, Cheng B, Zhang Y. Controlled dual delivery of angiogenin and curcumin by electrospun nanofibers for skin regeneration. Tissue Engineering. Part A. 2017;**23**(13-14):597-608. DOI: 10.1089/ten.tea.2016.0268

[148] Mengistu Lemma S, Bossard F, Rinaudo M. Preparation of pure and stable chitosan nanofibers by electrospinning in the presence of poly(ethylene oxide). International Journal of Molecular Sciences. 2016;**17**(11):E1790. DOI: 10.3390/ijms17111790

[149] Koosha M, Raoufi M, Moravvej H. One-pot reactive

electrospinning of chitosan/PVA hydrogel nanofibers reinforced by halloysite nanotubes with enhanced fibroblast cell attachment for skin tissue regeneration. Colloids and Surfaces. B, Biointerfaces. 2019;**179**:270-279. DOI: 10.1016/j.colsurfb.2019.03.054

[150] Mahmoudi N, Eslahi N, Mehdipour A, Mohammadi M, Akbari M, Samadikuchaksaraei A, et al. Temporary skin grafts based on hybrid graphene oxide-natural biopolymer nanofibers as effective wound healing substitutes: Pre-clinical and pathological studies in animal models. Journal of Materials Science. Materials in Medicine. 2017;**28**(5):73. DOI: 10.1007/s10856-017-5874-y

[151] Yousefi I, Pakravan M, Rahimi H, Bahador A, Farshadzadeh Z, Haririan I. An investigation of electrospun henna leaves extract-loaded chitosan based nanofibrous mats for skin tissue engineering. Materials Science & Engineering. C, Materials for Biological Applications. 2017;**75**:433-444. DOI: 10.1016/j.msec.2017.02.076

[152] Piran M, Vakilian S, Piran M, Mohammadi-Sangcheshmeh A, Hosseinzadeh S, Ardeshirylajimi A. In vitro fibroblast migration by sustained release of PDGF-BB loaded in chitosan nanoparticles incorporated in electrospun nanofibers for wound dressing applications. Artif Cells Nanomed Biotechnol. 2018;**46**(supp 1):511-520. DOI: 10.1080/21691401.2018.1430698

[153] Tanha S, Rafiee-Tehrani M, Abdollahi M, Vakilian S, Esmaili Z, Naraghi ZS, et al. G-CSF loaded nanofiber/nanoparticle composite coated with collagen promotes wound healing in vivo. Journal of Biomedical Materials Research. Part A. 2017;**105**(10): 2830-2842. DOI: 10.1002/jbm.a.36135

[154] Noh HK, Lee SW, Kim JM, Oh JE, Kim KH, Chung CP, et al. Electrospinning of chitin nanofibers: Degradation behavior and cellular response to normal human keratinocytes and fibroblasts. Biomaterials. 2006;**27**(21):3934-3944. DOI: 10.1016/j.biomaterials.2006.03.016

[155] Vashisth P, Pruthi V. Synthesis and characterization of crosslinked gellan/PVA nanofibers for tissue engineering application. Materials Science & Engineering. C, Materials for Biological Applications. 2016;**67**:304-312. DOI: 10.1016/j.msec.2016.05.049

[156] Vashisth P, Nikhil K, Roy P, Pruthi PA, Singh RP, Pruthi V. A novel gellan-PVA nanofibrous scaffold for skin tissue regeneration: Fabrication and characterization. Carbohydrate Polymers. 2016;**136**:851-859. DOI: 10.1016/j.carbpol.2015.09.113

[157] Vashisth P, Srivastava AK, Nagar H, Raghuwanshi N, Sharan S, Nikhil K, et al. Drug functionalized microbial polysaccharide based nanofibers as transdermal substitute. Nanomedicine. 2016;**12**(5):1375-1385. DOI: 10.1016/j.nano.2016.01.019

[158] Bonifacio MA, Cometa S, Cochis A, Gentile P, Ferreira AM, Azzimonti B, et al. Data on Manuka Honey/Gellan gum composite hydrogels for cartilage repair. Data in Brief. 2018;**20**:831-839. DOI: 10.1016/j.dib.2018.08.155

[159] Yu I, Kaonis S, Roland CR. A study on degradation behavior of 3D printed gellan gum scaffolds. Procedia CIRP. 2017;**65**:78-83. DOI: 10.1016/j.procir.2017.04.020

[160] Lin T, Lu C, Zhu L, Lu T. The biodegradation of zein in vitro and in vivo and its application in implants. AAPS PharmSciTech. 2011;**12**(1):172-176. DOI: 10.1208/s12249-010-9565-y

[161] Maharjan B, Joshi MK, Tiwari AP, Park CH, Kim CS. In-situ synthesis of AgNPs in the natural/synthetic hybrid

nanofibrous scaffolds: Fabrication, characterization and antimicrobial activities. Journal of the Mechanical Behavior of Biomedical Materials. 2017;**65**:66-76. DOI: 10.1016/j.jmbbm.2016.07.034

[162] Zhang M, Li X, Li S, Liu Y, Hao L. Electrospun poly(l-lactide)/zein nanofiber mats loaded with Rana chensinensis skin peptides for wound dressing. Journal of Materials Science. Materials in Medicine. 2016;**27**(9):136. DOI: 10.1007/s10856-016-5749-7

[163] Figueira DR, Miguel SP, de Sá KD, Correia IJ. Production and characterization of polycaprolactone-hyaluronic acid/chitosan- zein electrospun bilayer nanofibrous membrane for tissue regeneration. International Journal of Biological Macromolecules. 2016;**93**(Pt A): 1100-1110. DOI: 10.1016/j.ijbiomac.2016.09.080

[164] Babitha S, Korrapati PS. Biodegradable zein-polydopamine polymeric scaffold impregnated with TiO2 nanoparticles for skin tissue engineering. Biomedical Materials. 2017;**12**(5):055008. DOI: 10.1088/1748-605X/aa7d5a

[165] Sultana N, Khan TH. In vitro degradation of PHBV scaffolds and nHA/PHBV composite scaffolds containing hydroxyapatite nanoparticles for bone tissue engineering. Journal of Nanomaterials. 2012;**2012**:190950. DOI: 10.1155/2012/190950

[166] Sundaramurthi D, Krishnan UM, Sethuraman S. Epidermal differentiation of stem cells on poly(3-hydroxybutyrate-co-3-hydroxyvalerate) (PHBV) nanofibers. Annals of Biomedical Engineering. 2014;**42**(12):2589-2599. DOI: 10.1007/s10439-014-1124-3

[167] Keshel SH, Biazar E, Rezaei Tavirani M, Rahmati Roodsari M, Ronaghi A, Ebrahimi M, et al. The healing effect of unrestricted somatic stem cells loaded in collagen-modified nanofibrous PHBV scaffold on full-thickness skin defects. Artif Cells Nanomed Biotechnol. 2014;**42**(3):210-216. DOI: 10.3109/21691401.2013.800080

[168] Zine R, Sinha M. Nanofibrous poly(3-hydroxybutyrate-co-3-hydroxyvalerate)/collagen/graphene oxide scaffolds for wound coverage. Materials Science & Engineering. C, Materials for Biological Applications. 2017;**80**:129-134. DOI: 10.1016/j.msec.2017.05.138

[169] Yuan J, Geng J, Xing Z, Shim KJ, Han I, Kim J, et al. Novel wound dressing based on nanofibrous PHBV-keratin mats. Journal of Tissue Engineering and Regenerative Medicine. 2015;**9**(9):1027-1035. DOI: 10.1002/term.1653

Section 4

Microfluidic Devices

Chapter 6

Microfluidic Device for Single Cell Impedance Characterization

Muhammad Asraf Mansor and Mohd Ridzuan Ahmad

Abstract

Detection of single particle has emerged as a noninvasive technique for diagnostic and prognostic patients with cancer suspected. Microfluidic impedance cytometry has been utilized to detect and measure the electrical impedance of single biological particles at high speed. The detailed information of single cells such as cell size, membrane capacitance, and cytoplasm conductivity also can be obtained by impedance measurement over a wide frequency range. In this work, we developed an integrated microneedle microfluidic device to detect and discriminate 9- and 16-μm microbeads. Two microneedles were utilized as measuring electrodes at the half height of the microfluidic device to perform measurement of electrical impedance under a presence of cells at the sensing area. Furthermore, this device was able to distinguish the cell concentration in the suspension fluid. The reusable microneedles were easy to be inserted and withdrawn from the disposable microfluidic. The ultrasonic cleaning machine has been used to clean the reusable microneedle with a simple cleaning process. Despite of the low-cost device, its capability to detect single particles at the sensing area was preserved. Therefore, this device is suitable for cost-efficient medical and food safety screening and testing process in developing countries.

Keywords: impedance, flow cytometry, microfluidics, microneedle, single cell

1. Introduction

The single cell analysis (SCA) has been emphasized to provide biologists and scientists to peer into the molecular machinery of individual cells. For the application of medical diagnosis, detection of cancer cells and pathogenic bacteria cells in blood is utilized as a diagnosing infectious disease. It is reported that detection of circulating tumor cells (CTCs) in the blood has shown to be clinically important for early stage metastasis or recurrence of cancer. The presence of rare CTCs in blood is ranging from only 1–100 CTCs/ml blood [1]. *Plasmodium falciparum* malaria, which kills mainly children in developing countries infected the blood sample of patients at concentration of ~1/50 μl of blood [2]. Nowadays, the analysis of single cell in biological measurements and medical research has emerged as a distinct new field and acknowledged to be one of the fundamental building blocks of life [3]. Amongst of various single cell analysis, cell impedance measurement has become an effective method of biological measurement [4]. The physiological behavior of the cells and their corresponding molecular expressions have significant effect on the cell membrane and cytoplasm conductivity and dielectric constant, which in turn affects the overall impedance characteristics [5]. For that reason, the impedance measurements on single cells can provide relevant information about its functional status and may be a simple and significantly less complex alternative to detailed molecular expression studies.

The classical method for cell detection in suspension is using flow cytometry, which is rapid and highly accurate measurement technique. Impedance flow cytometry is an indirect signal extraction from the single cells on microchannel sensing area without having direct access into intracellular region of the cells [6]. These techniques were first reported by Coulter [7] has emerged in the microfabrication device in order to analyze microscale particle with high sensitivity. However, flow cytometry involves expensive manufacturing and labelling of the cells with fluorescent antibodies [8]. Recently, the impedance flow cytometry (IFC) has gained attention for the significant promising techniques to replace and overcome the limitations associated with flow cytometry. The IFC is preferable because of fast, real-time, and non-invasive methods for biological detection. This technique is capable to be utilized as cell counting [8], cancer cell detection [9] and bacteria detection [10]. Some groups have demonstrated detection and counting of cells by using a microfluidic integrated with electrode for various electrical measurement methods in an application of food safety [11] and real-time monitoring bio-threat [12]. This measurement technique is based on the alteration of impedance across a measurement electrode due to the blocking of ionic current passing between electrodes when a presence of the cells.

The IFC is capable to distinguish and count lymphocytes, monocytes and neutrophils in human whole blood [8]. Other studies reported that IFC can detect the presence of cells based on probing the impedance inside the cell at frequency greater than 1 MHz [13]. Fabricated nanoneedles probe inside microfluidic was utilized for measuring the presence of cells at sensor surface and making it sensitive to the dielectric properties of solution [14]. However, this device requires patterning of electrode or probe on the substrate resulting in higher cost of the fabrication process. Another limitation also needs to consider is time-consuming cleaning process of the device. Several groups have demonstrated the technique to reduce the cost of microfabrication of electrodes by using printed circuit board (PCB) as a measurement electrode. They demonstrated contactless conductivity detection in capillary electrophoresis manners [15] and cell manipulation using dielectophoresis [16]. Recently, the contactless impedance cytometry was developed to reduce the fabrication cost of impedance cytometry device [17, 18]. The electrode was fabricated on the PCB substrate (reusable component) and the thin bare dielectric substrate bonded to a PDMS microchannel (disposable component) was placed onto PCB substrate. The sensitivity of this device is the limitation since the electric field was buried in dielectric substrate and not reaches the electrolyte. Several designs and method in IFC in order to detect and analyze a cell have been reported [19, 20].

This chapter discusses a novel integrated microneedles-microfluidic system for detecting yeast cell concentration in suspension as well as detecting a single particle based on the impedance measurement. The development of the device focuses on reducing the fabrication cost while preserving the main functionality, that is, cell detection. The significant fabrication cost reduction in this work is by replacing the microfabrication of electrodes by the microneedles. This device utilized a Tungsten microneedle as a measurement electrode which can be reused and easily to be cleaned. The two microneedles were placed at half height disposable microchannel to detect and enable impedance measurement of passing cells through the applied electric field. **Figure 1(a)** illustrated the schematic diagram of the proposed microfluidic chip which consists of two microneedles integrated at both sides of the microchannel. The main sensing area microchannel length, width and thickness are

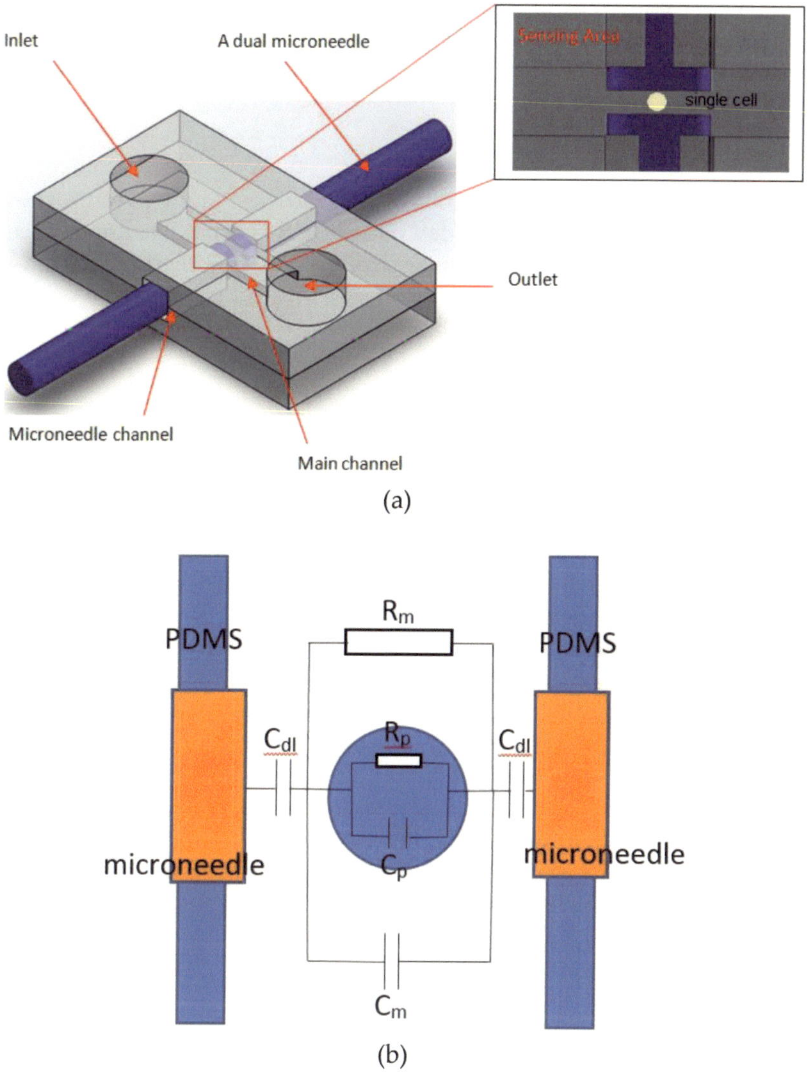

Figure 1.
(a) 3D schematic diagram of the microfluidic device structure integrated with microneedle and top view of sensing area which the impedance measurement of single particle be measured. (b) Microfluidic sensing area equivalent circuit model.

100, 25 and 25 μm respectively. The device is suitable for early cancer cell detection application in developing countries since it significantly reduces the fabrication cost, that is, not required the fabrication of micro electrode.

1.1 Principle

Ohm's Law has been use as the basic principle of detecting suspended biological cells in the media. The passing cells across the sensing area have been measure by an AC current with a frequency sweep to determine the changing impedance value of media. **Figure 1(b)** illustrated the equivalent circuit model for obtain the all parameter involved in order to characterize the electrical properties of cell between two microneedles in the suspension media. The sensing area of mirofluidic chip can be modeled electrically as cell impedance of resistance R_p and cell capacitance C_p in parallel with the impedance contributed by all materials between the two electrodes, which consist solution resistor R_m in parallel with solution capacitance C_m. Both impedance in series with a pair of electrodes capacitance double layer C_{dl}. Z_T is overall impedance of the measurement system given [21].

$$Z_T = \frac{2}{j\omega C_{dl}} + \frac{R_m R_p}{R_m + R_p + j\omega R_m R_p (C_m + C_p)} \tag{1}$$

where ω is the angular frequency of the electrical signal. As a result, the Z_T is changing according the present of cell in the sensing area. The impedance between electrode (microneedle) and electrolyte (solution medium) is our main focus in this work.

1.2 Experimental works

1.2.1 Cell culture

In this work, Sacharomycesceresiae cells and microparticle are used as a model for proof of concepts. Sacharomycesceresiae were cultivated in a petri dish containing 10 ml of YPD broth (Yeast extract Peptone Dextrose). The YPD broth contained 1% yeast extract, 2% peptone and 2% glucose. The YPD dishes were incubated at 37°C for 24 hours. The cells were washed with deioinsed (DI) water three times by centrifugation, then they were suspended in sterilized deioinsed water at various dilutions (1:10) concentration. The cells were incubated on agar plates at 37°C for 24 hours for determining the number of cells. The diameter of yeast cells varies from 4 to 7 μm. The number of cells was 1.3×10^9 colony forming units per milliliter (cfu/ml). The conductivity of DI water is 6 mS/m. The non-fluorescent polystyrene (PS) microbeads with diameter 15 and 9 μm (Polysciences. Inc.) suspended in Phosphate-buffered saline (PBS) solution were diluted to a final concentration of 1000 beads per ml. Polystyrene beads have a known size and electrical properties [22] and have constant impedance across the frequency range used in these experiments.

1.2.2 Device fabrication

The photolithography technique was utilized to fabricate the microfluidic device. The fabrication begins by designing the masks using layout editor software. The laser lithography system (μPG501, Heidelberg Instruments, Germany) has been used to write the two masks (top and bottom) on the chromium (Cr) masks. Two-step photolithography using SU-82025 negative photoresist (MicroChem, USA) was utilized to fabricate the top layer mold. The first layer has a thickness of 25 μm and was spin coated onto a silicon substrate. After pre-baking, the top layer of Cr mask was place onto the first layer of photoresist for pattern transfer by using a mask aligner (SussMicroTech MA-6), then post-baking with development. Next, the second layer with 60 μm thickness was spin coated on the first photoresist layer and pre-baking. Then, the second layer of photoresist substrate was exposed with the bottom layer Cr mask by UV light. After exposed, the top mold master was obtaining by post-bake and developed process. Meanwhile the bottom mold master with 60 μm thickness was fabricated by following the SU-8 microchannel photolithography technique. PDMS pre-polymers (SYLGARD184A) was thoroughly mixing with curing agents (SYLGARD 184B) in a ratio of 10:1 by weigh for fabricate the PDMS microfluidic chip. The mixing PDMS was poured on an SU-8 mold master (top and bottom mold master) and left for whole night cured at room temperature to obtain the PDMS microfluidic chip. To increase the bonding strength between the top side and bottom side of PDMS microfluidic chip, they were cleaned with Isopropyl alcohol (IPA) and treated by Oxygen plasma (Plasma Etch PE-25) for 25 seconds [23]. The bonding process of both side PDMS microfluidic chip was completed in less than 2 minutes to prevent loss of Oxygen plasma effectiveness. Finally, the right and left sides of the microchannel chip were cut and leaving a square (60 μm × 120 μm) hole for inserting a microneedle. For measuring electrode, two Tungsten microneedle (Signatone) coated by parylene with tip diameter, shank diameter and length of tungsten needle are 25, 250 and 31.7 mm, respectively, was utilized.

1.2.3 Device operation

The microscope (Olympus Inverted Microscopes IX71) was utilized to monitor the sensing area of microfluidic chip system. The micromanipulator (EB-700, Everbeing) was utilized to insert the two microneedles into microchannel chip through the square hole at right and left side of the chip. For this experiment, the gap between microneedles was fixed at 20 μm. **Figure 2** illustrated the experimental setup of impedance measurement. The two microneedles were connected to impedance analyzer (Hioki IM3570) for input measuring and the result was displayed on the computer. Then, by controlling the syringe pumps (KDS LEGATO 111, KD Scientific, and USA), the 3-ml syringes of the sample solution and yeast concentration was introduced into microfluidic chip via two tygon flexible tubes.

1.2.4 Impedance measurement procedure

Standard short and open self-calibration procedure has been used for impedance analyzer in order to perform the impedance measurement. Furthermore, to calibrate the chip, impedance of 1xPBS solution was measured at the 20 μm of electrode gap. Three microfluidic devices were utilized for reproducibility testing and the experiments were conducted at room temperature. To validate the equivalent circuit model, impedance of the medium between microneedle was measured. The solutions were sterilized DI water and PBS (Photosphate-buffered saline) with conductivities 6 and 1.4 S/m respectively.

Initially, 1 ml of PBS with concentrations of 1500 mOsm was prepared for the chip cleaning process. The sample was loaded into a syringe and driven through the microchannel using a syringe pump with the flow rate of syringe pump was kept constant (60 μl/min). After flushing with PBS solution, yeast cell of 1 ml of each seven different concentrations of sample from 10^2 to 10^9cfu/ml were driven

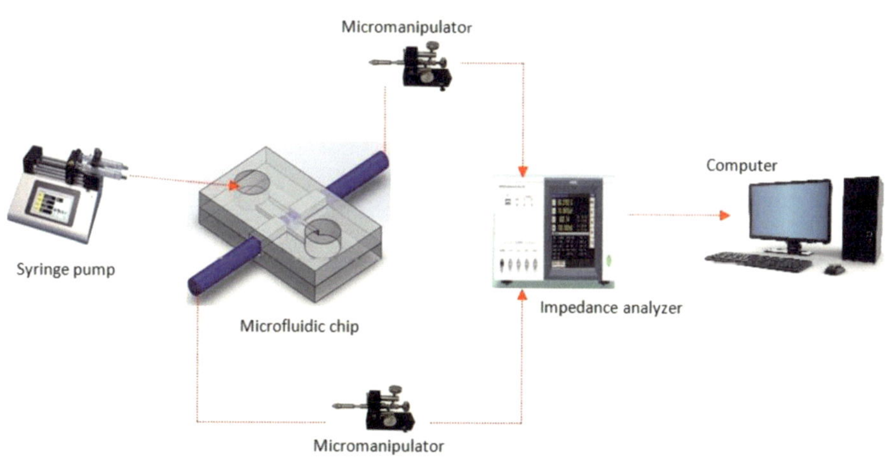

Figure 2.
The experimental set-up diagram.

through the microchannel at a flow rate of 6 μl/min. The impedance of each solution was measure by connecting the microneedles with impedance analyzer. Then, an AC signal frequency range from 100 Hz to 5 MHz with an applied voltage of 1 V was apply to determine the impedance spectra (impedance and phase vs. frequency) in order to differentiate the variations of solution samples. Between each sample measurement, the microchannel chip was flushed by DI and PBS water for 1 and 2 minutes respectively. The impedance analyzer (Hioki IM3570) GUI and post-processed in MATLAB (MathWorksInc, USA) was utilized to record the data. The impedance change during the passage yeast cells at sensing area was measured. In order to monitor the behavior of impedance for each sample, the impedance value at three frequencies (100 kHz, 500 kHz and 1 MHz) was measured.

Single cell detection and measurement was conducted based on impedance measurement with or without single cell at the sensing area. Two sample of microbeads with diameter 15 and 9 μm suspended in 1 ml of PBS with concentration of 10^3 per ml were utilized to perform this measurement and detection. Each sample were driven through the microchannel at a flow rate of 6 μl/min and measured using an AC signal frequency range from 100 Hz to 5 MHz.

1.3 Result and discussion

As a proof of concepts, the dependencies of the impedance on the various concentrations of yeast cells and a single microbead in the suspension medium by using this microfluidic device are studied. **Figure 3** presents the measured impedance spectra and fitting spectra (on a log scale) of the system for two of microchannel filled with sterilized DI water and PBS at frequency range 1 kHz to 1 MHz. For simulation, 100 data points on the impedance measured spectrum were used as input to the equivalent circuit [see **Figure 1(b)**] and generating the fitting impedance spectrum by using MATLAB. For high conductivity fluid (PBS), the result shows two domains which were an electrical double layer (EDL) region and a resistive region [24]. The EDL occurred in the low frequency range from 1 kHz to approximately 300 kHz, whereas the resistive region occurred in high frequency from 300 kHz to 1 MHz. The agreement between the measured and fitting spectra result indicated that our developed circuit model for this system is feasible to determine the impedance characteristics of solution medium.

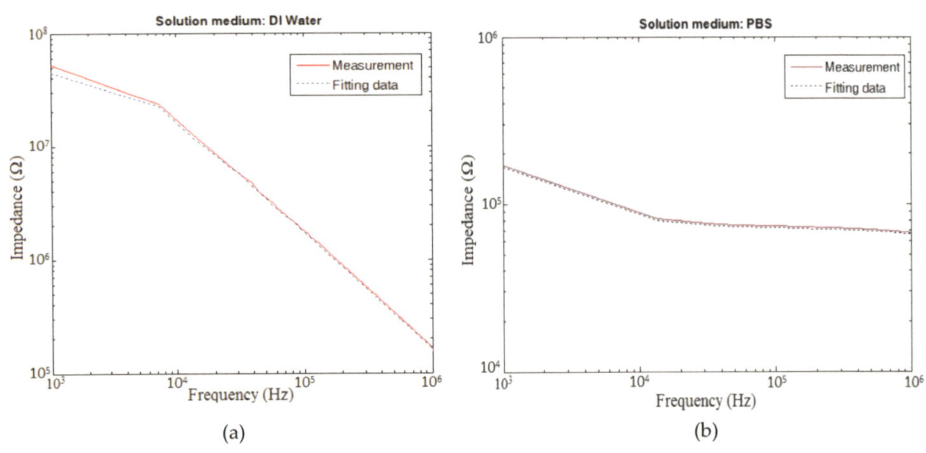

Figure 3.
Impedance spectra of sample solution together with their fitting spectra: (a) DI water (b) PBS.

To illustrate the cell detection capability of the device, yeast cell and microbeads with different concentration was utilized. Yeast cells concentration ranging from 10^2 to 10^9 cfu/ml were infused inside microchannel with fixed flow rate 6 μl/min and fixed electrode gap (25 μm). A sweep frequency (100 kHz to 5 MHz) AC signal (1 V) was applied to the one side of the microneedle and the current entering at another side of microneedle was measured to calculate the impedance of concentration of yeast cells in DI water. Initially, 10^9 cfu/ml was injected resulting in a drop-in impedance by referring the impedance of DI water as a control. Afterward, the microchannel chip was washed by the PBS followed by DI water at maximum flow rate.

The maximum flow rate of the liquid can flow inside microchannel without leaking is 300 μl/min. **Figure 4(a)** shows the impedance spectra of yeast cell in DI water with the different cell concentration in the range 10^4–10^9cfu/ml, along with DI water as a reference. After washing the microchannel, 10^8cfu/ml was infused to the microchannel resulting in an increase in impedance. It can be seen the impedance spectra of yeast cell in DI water across the sensing area (two microneedles)

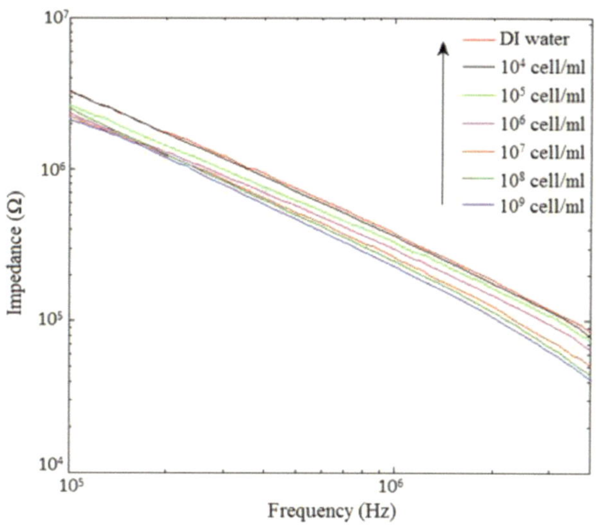

(a)

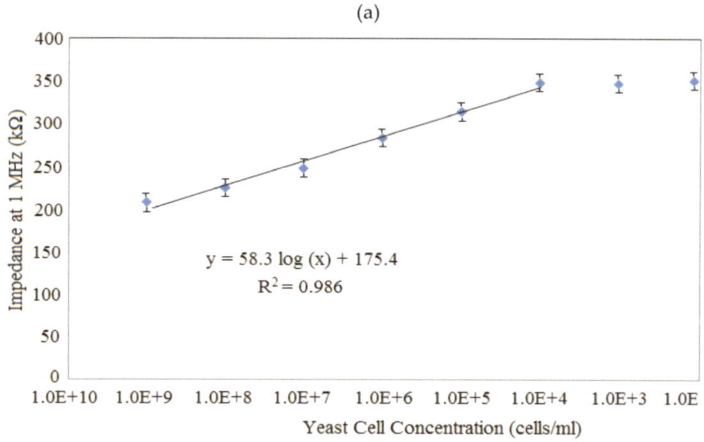

(b)

Figure 4.
(a) Impedance spectra of yeast cells in water with cell concentrations ranging from 10^2 to 10^9cfu/ml, along with DI water as controls; (b) the logarithmic value of the concentration of yeast cells and the impedance measured at 1 MHz with linear relationship.

increase with decreasing the cell concentration of cells [14]. According to the observation result, it can be said that cell suspension with high concentrations is more conductive than those with lower concentrations.

The conductivity of solution varies proportionally to the number of cell concentration at fixed volume of solution [25]. In some studies, the relative dielectric permittivity and charged polyelectrolytes inside the cell also may affect the impedance of solution [14]. The optimum region for sensing microneedle to differentiate the cell concentration in DI water occurs between 500 kHz to 5 MHz. The impedance values of the suspensions in this frequency region were significantly different from each other. The experiment was repeated two times measurement cycle and showed the similar result.

In cell detection experiment, frequency lower than 100 kHz are not considered, as the EDL started to influence the measurement at frequency below 300 kHz [17, 26]. In order to investigate the relationship between impedance value and cell concentration, we selected 1 MHz as the best representative frequency. **Figure 4(b)** illustrates the impedance responses of the sample containing different yeast cell concentrations and DI water at frequency measurement 1 MHz. The impedance of the solution was significantly increased from 207.63 to 225.42, 247.61, 284.48, 314.64, and 348.51 kΩ when the yeast concentration decreasing from 10^9 to 10^8, 10^7, 10, 10 and 10^4 cfu/ml respectively. After the cell concentrations were lower than 10^4 cfu/ml, impedance value shows no significant changing between each other or DI water. In additions, the pattern of the result shows a linear relationship between the impedance and the logarithmic value of the cell concentration at cell concentration from 10^4 to 10^9 cfu/ml (see **Figure 4**). The linear regression equation of this result is Z (kΩ) = 58.3 log X (cells/ml) + 175.4 with R2 = 0.986. The detection limit was calculated to be 1.2×10^4 cfu/ml. Error bars are standard deviations of five measurements cycle.

In order to measure the cell concentration in DI water suspensions, the linear regression equation of the impedance of the yeast suspensions was used. This device can be utilized to quantify cells in suspensions other than impedance microbiology and impedance biosensors for bacteria detection since the detection limit of this method is comparable with another sensor. The reported sensor for detection of pathogenic bacteria are QCM immunosensors for detection of Salmonella with detection limits of 9.9×10^5 cfu/ml [27], surface plasmon resonance (SPR) sensor for detection of *E. coli* O157:H7 with a detection limit of 10^7 cfu/ml [28] and SPR immounosensors for detection of Salmonella enteritidis and Listeria monocytogens with detection limits of 10^6 cfu/ml [29].

In order to demonstrate the capability of this device in detecting the present of single cell, two size of micro bead have been flowing inside the microfluidic device. The impedance of PBS solution as a control was initially infused inside the microfluidic device. Then two samples of microbeads in PBS solution were infused inside the microchannel with the same flow rate and electrode gap of yeast cell concentration measurement. **Figure 5(a)** shows the 15-μm microbead flow through the sensing area and a sweep frequency ranging from 100 kHz to 3 MHz AC signal (1 V) was applied to the electrode. As the result, the impedance spectrum is plotted over the field frequency, as shown in **Figure 5(b)**. The figure shows the average electrical impedance data for two size of beads and PBS solution without present of beads. From this average data, it is expected that the electrical impedance spectrum can be used to differentiate between the sizes of beads. The beads (9 and 15 μm in diameter) are clearly discriminated by impedance spectrum. The impedance increases with the increasing of the size of particle. Due to the presence of a single bead that can be regarded as an insulating object, the electrical resistance of the sensing channel was slightly increased.

As the result, we conclude this device can detect the cell concentrations in solution medium and the single microbead at the high frequency range between 100 kHz and 5 MHz. In this experiment, we did not determine the detection capability at the frequency lower than 100 kHz. For the future work, we will focus on the measurement and detection to the human cell (normal and cancer cell) the size of microneedle, single cell detection and utilize a non-polarizable electrode, that is, Ag/AgCl (to eliminate the EDL), in order to improve the performance of the device.

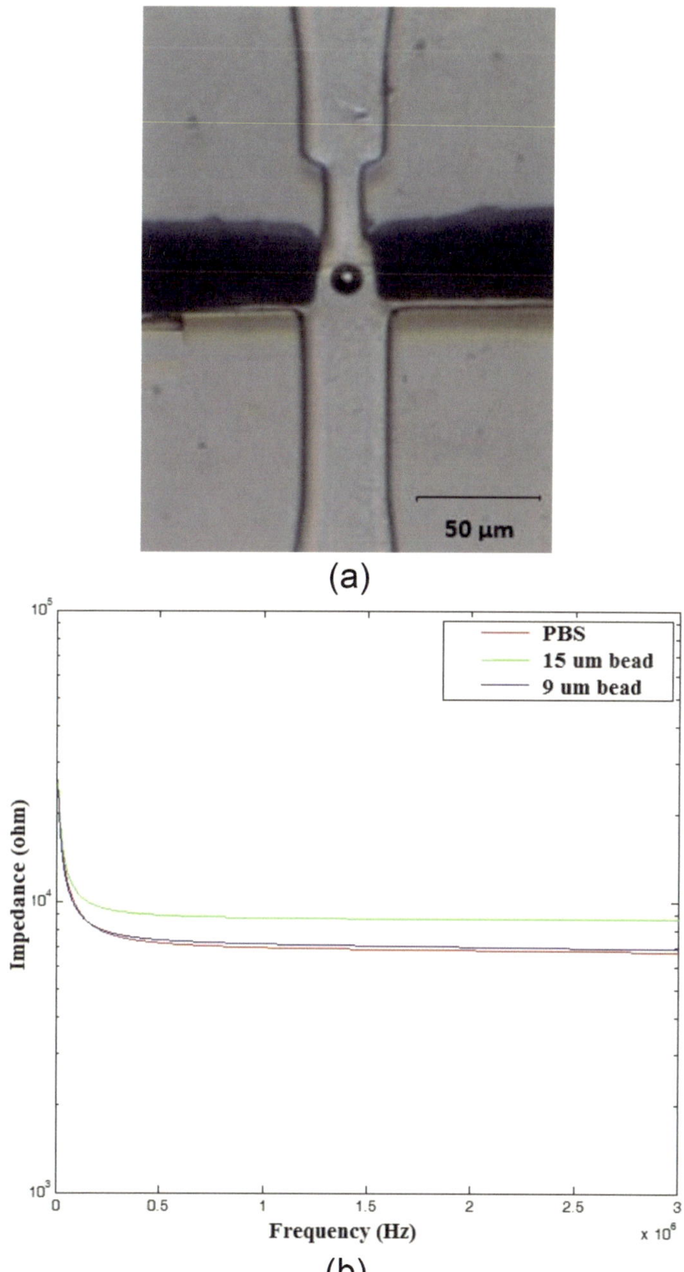

Figure 5.
(a) The single microbead with diameter of 15 µm flows through the sensing area. (b) Impedance spectrum of two different sizes of beads in PBS solution and PBS solution (without bead).

2. Conclusions

For conclusion, a simple, low-cost and label-free microfluidic device has developed to detect the cell concentration and single cell in the suspension medium. This device contains a disposable PDMS microchannel which allow a reusable microneedle insert into microchannel. The result demonstrated the increase of cell concentration in the solution medium were decrease the impedance value. The capability of this device to differentiate the concentration of cell from 10^9 to 10^4 cfu/ml shows the core functionality of the proposed sensor even though the manufacturing cost was significantly lower. In addition, the microfluidic device capable to detect single cell and decimate the size of single cell. As a proof of concept, yeast cell and microbeads were used in this study and we emphasize this sensing technique can be applied to a variety of cell types with diameter size in a range from 5 to 25 μm. It is recommended to perform only one measurement time for each PDMS microchip in order to avoid the potential spread of contamination to samples. The fabrication cost of this device is significantly reduces ($\approx$ 30% fabrication cost was reduced based on facility rental and raw material usage) which is suitable for early cancer cell detection and water contamination application in developing countries.

Acknowledgement

The research was supported by the Ministry of Higher Education of Malaysia and Universiti Teknologi Malaysia (Grant Nos. Q.J130000.21A2.03E11 and Q.J130000.2423.03G47); we thank them for funding this project and for their endless support.

Author details

Muhammad Asraf Mansor and Mohd Ridzuan Ahmad*
Micro-Nano System Engineering Research Group, Faculty of Engineering, Division of Control and Mechatronic Engineering, School of Electrical Engineering, Universiti Teknologi Malaysia, Skudai, Johor, Malaysia

*Address all correspondence to: mdridzuan@utm.my

References

[1] Kantara C et al. Methods for detecting circulating cancer stem cells (CCSCs) as a novel approach for diagnosis of colon cancer relapse/metastasis. Laboratory Investigation. 2015;**95**(1):100-112

[2] Ciceron L, Jaureguiberry G, Gay F, Danis M. Development of a plasmodium PCR for monitoring efficacy of antimalarial treatment. Journal of Clinical Microbiology. 1999;**37**(1):35-38

[3] Gilchrist KH et al. General purpose, field-portable cell-based biosensor platform. Biosensors & Bioelectronics. 2001;**16**(7-8):557-564

[4] Jao J-Y, Liu C-F, Chen M-K, Chuang Y-C, Jang L-S. Electrical characterization of single cell in microfluidic device. Microelectronics and Reliability. 2011;**51**(4):781-789

[5] Yang L, Arias LR, Lane TS, Yancey MD, Mamouni J. Real-time electrical impedance-based measurement to distinguish oral cancer cells and non-cancer oral epithelial cells. Analytical and Bioanalytical Chemistry. 2011;**399**(5):1823-1833

[6] Sun T, Morgan H. Single-cell microfluidic impedance cytometry: A review. Microfluidics and Nanofluidics. 2010;**8**(4):423-443

[7] Coulter WH. High speed automatic blood cell counter and cell analyzer. Proceedings of the National Electronic Conference. 1956;**12**:1034-1040

[8] Holmes D, Morgan H. Single cell impedance cytometry for identification and counting of CD4 T-cells in human blood using impedance labels. Analytical Chemistry. 2010;**82**(4):1455-1461

[9] Gou H-L, Zhang X-B, Bao N, Xu J-J, Xia X-H, Chen H-Y. Label-free electrical discrimination of cells at normal, apoptotic and necrotic status with a microfluidic device. Journal of Chromatography. A. 2011;**1218**(33):5725-5729

[10] Du E, Ha S, Diez-Silva M, Dao M, Suresh S, Chandrakasan AP. Electric impedance microflow cytometry for characterization of cell disease states. Lab on a Chip. 2013;**13**(19):3903-3909

[11] Liu Y-S, Banada PP, Bhattacharya S, Bhunia AK, Bashir R. Electrical characterization of DNA molecules in solution using impedance measurements. Applied Physics Letters. 2008;**92**(14):143902

[12] Javanmard M, Talasaz AH, Nemat-Gorgani M, Pease F, Ronaghi M, Davis RW. Targeted cell detection based on microchannel gating. Biomicrofluidics. 2007;**1**(4):1-10

[13] Gawad S, Schild L, Renaud PH. Micromachined impedance spectroscopy flow cytometer for cell analysis and particle sizing. Lab on a Chip. 2001;**1**(1):76-82

[14] Esfandyarpour R, Javanmard M, Koochak Z, Harris JS, Davis RW. Nanoelectronic impedance detection of target cells. Biotechnology and Bioengineering. 2014;**111**(6):1161-1169

[15] Wang J, Chatrathi MP, Mulchandani A, Chen W. Capillary electrophoresis microchips for separation and detection of organophosphate nerve agents. Analytical Chemistry. 2001;**73**(8):1804-1808

[16] Park K, Suk H-J, Akin D, Bashir R. Dielectrophoresis-based cell manipulation using electrodes on a reusable printed circuit board. Lab on a Chip. 2009;**9**(15):2224-2229

[17] Emaminejad S, Javanmard M, Dutton RW, Davis RW. Microfluidic

diagnostic tool for the developing world: Contactless impedance flow cytometry. Lab on a Chip. 2012;**12**(21):4499

[18] Emaminejad S, Paik K, Tabard-Cossa V, Javanmard M. Portable cytometry using microscale electronic sensing. Sensors and Actuators B: Chemical. 2016;**224**:275-281

[19] Mansor MA, Ahmad MR. Single cell electrical characterization techniques. International Journal of Molecular Sciences. 2015;**16**(6):12686-12712

[20] Mansor MA, Ahmad MR. A simulation study of single cell inside an integrated dual Nanoneedle-microfluidic system. Jurnal Teknologi. 2016;**78**(7-5):59-65

[21] Sun T, Holmes D, Gawad S, Green NG, Morgan H. High speed multi-frequency impedance analysis of single particles in a microfluidic cytometer using maximum length sequences. Lab on a Chip. 2007;**7**(8):1034-1040

[22] Spencer D, Hollis V, Morgan H. Microfluidic impedance cytometry of tumour cells in blood. Biomicrofluidics. 2014;**8**(6):064124

[23] Mansor M, Takeuchi M, Nakajima M, Hasegawa Y, Ahmad M. Electrical impedance spectroscopy for detection of cells in suspensions using microfluidic device with integrated microneedles. Applied Sciences. 2017;**7**(2):170

[24] Morgan H, Sun T, Holmes D, Gawad S, Green NG. Single cell dielectric spectroscopy. Journal of Physics D: Applied Physics. 2007;**40**:61-70

[25] Yang L. Electrical impedance spectroscopy for detection of bacterial cells in suspensions using interdigitated microelectrodes. Talanta. 2008;**74**(5):1621-1629

[26] Segerink LI, Sprenkels AJ, ter Braak PM, Vermes I, van den Berg A. On-chip determination of spermatozoa concentration using electrical impedance measurements. Lab on a Chip. 2010;**10**(8):1018

[27] Park IS, Kim WY, Kim N. Operational characteristics of an antibody-immobilized QCM system detecting salmonella spp. Biosensors & Bioelectronics. 2000;**15**(3-4):167-172

[28] Fratamico PM, Strobaugh TP, Medina MB, Gehring AG. Detection of Escherichia coli 0157:H7 using a surface plasmon resonance biosensor. Biotechnology Techniques. 1998;**12**(7):571-576

[29] Koubová V et al. Detection of foodborne pathogens using surface plasmon resonance biosensors. Sensors and Actuators B: Chemical. 2001;**74**(1-3):100-105

Chapter 7

Assessing the Vascular Deformability of Erythrocytes and Leukocytes: From Micropipettes to Microfluidics

Mark D. Scott, Kerryn Matthews and Hongshen Ma

Abstract

Among the most crucial rheological characteristics of blood cells within the vasculature is their ability to undergo the shape change (i.e., deform). The significance of cellular deformability is readily apparent based solely on the disparate mean size of human erythrocytes (~8 μm) and leukocytes (10–25 μm) compared to the minimum luminal size of capillaries (4–5 μm) and splenic interendothelial clefts (0.5–1.0 μm) they must transit. Changes in the deformability of either cell will result in their premature mechanical clearance as well as an enhanced possibility of intravascular lysis. In this chapter, we will demonstrate how microfluidic devices can be used to examine the vascular deformability of erythrocytes and agranular leukocytes. Moreover, we will compare microfluidic assays with previous studies utilizing micropipettes, ektacytometry and micropore cell transit times. As will be discussed, microfluidics-based devices offer a low-cost, high throughput alternative to these previous, and now rather ancient, technologies.

Keywords: deformability, hemorheology, red blood cells, white blood cells, micropipette assay, ektacytometry, cell transit analysis, microfluidic analysis, transfusion medicine

1. Introduction

The circulating cellular elements of blood consist of erythrocytes (red blood cells; RBC), leukocytes (white blood cells; WBC) and platelets. The hemorheology of these blood cells is unique in that these cells exist in a fluid phase subjected to variable, and often extreme, rheological shear stress, viscosity changes and biomechanical obstacles (e.g., capillaries and splenic filtration). Hemodynamically, shear stress is induced by the highly variable flow rate of blood within the ~100,000 kilometers of the human vasculature bed which encompasses both large arteries and veins to the capillary beds (**Figure 1A**) [1]. With an average resting cardiac output of approximately 5 L/min, blood flow in the largest artery (i.e., aorta) is approximately 50 cm/s while flow rates drop to only about 0.03 cm/s in the smallest capillaries and return to about 15–40 cm/s in the largest veins (e.g., superior and inferior vena cava) [1, 2]. In high flow conditions, RBC reside in the fast flowing central axial column of the vessel while WBC (and platelets) are located more peripherally

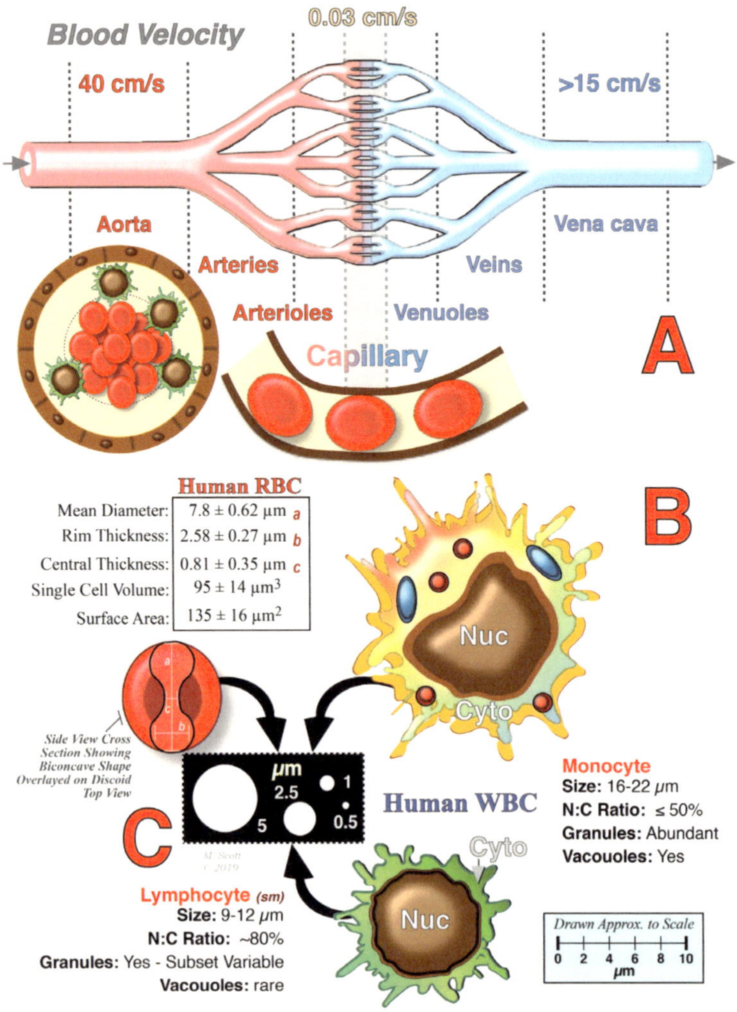

Figure 1.
The physiology and morphology of the vascular bed and blood cells imparts unique rheological stress on circulating blood cells. Panel A: the vascular bed is composed of blood vessels of various sizes which create significant disparity in blood (fluid and cellular) velocity consequent to vessel diameter. The fluid flow induces rheological sheer stress while the vessel size can create biomechanical deformation of cellular elements. Panel B: shown are the general physical parameters of human RBC and WBC. Note that the biconcave RBC is a anuclear while within the WBC, the nucleus:cytoplasm (N:C) ratio of monocytes and lymphocytes are quite divergent. RBC cytoplasmic viscosity is primarily defined by hemoglobin while in WBC, in addition to the nucleus, the presence of granules and vacuoles also impact intracellular viscosity and the aggregate cellular deformability. Panel C: blood cell deformability is crucial during vascular flow due to the size disparity between red blood cells and various leukocytes (e.g., monocytes and lymphocytes) and the capillary (4–58 μm) and splenic interendothelial clefts (0.5 μm). Panels B,C are drawn approximately to scale.

and prone to mechanical interaction with the endothelial cells lining the blood vessels. WBC also have adhesion molecules on their membrane and, if appropriate signals (e.g., inflammation) are present, they actively roll on the endothelial cells prior to attachment and extravasation (**Figure 1A,B**). Moreover, the viscosity of blood is also variable and is a function of, primarily, red blood cell (RBC) number and flow rate. At high RBC counts and high flow rates, blood is highly viscous while at low RBC counts and low flow rates (capillaries), blood viscosity is greatly reduced. Moreover, as shown in **Figure 1C**, the rheological stress is further exacerbated by the biomechanical stresses induced by the extreme disparity in the size of RBC (~8 μm) and WBC (10–25 μm) to the minimum diameter of the vascular capillary beds (4–5 μm) and splenic interendothelial clefts (0.5–1.0 μm) [3, 4]. Hence, consequent to both the shear forces, viscosity and biomechanical stresses placed on blood cells, a key biologic/physiologic requirement of both RBC and WBC within the vascular space is rheological deformability. Biomechanically, the intracellular viscosity and membrane rigidity of the RBC and WBC are the key factors in imparting their vascular rheological deformability.

For the anuclear RBC, intracellular viscosity is primarily determined by hemoglobin content (both absolute content and hemoglobin structure (**Figure 1B**)). RBC membrane deformability/flexibility is primarily imparted by the cytoskeletal structure of the cells and, to a lesser extent, the composition of the bilayer itself (lipid species, protein content, integral versus peripheral membrane proteins, and carbohydrates). For normal RBC the intra- and inter-individual variability of both intracellular viscosity is relatively invariant; however, genetic mutations affecting hemoglobin structure (e.g., HbS, α and β thalassemia, HbE mutations) will dramatically affect both hemoglobin content and the viscosity of the hemoglobin itself. Similarly, the cytoskeletal structure of normal red blood cells is both well characterized and consistent within humans. But, as with hemoglobin variants, mutations in any component of the cytoskeleton can dramatically affect the discoid shape of the RBC and result in size changes and/or altered rigidity or stability of the cytoskeleton and cell itself. Indeed, numerous studies have documented that changes in either the hemoglobin content or structure (the major determinant of viscosity) or mutations to cytoskeletal components (the major determinant of membrane rigidity) can exert significant effects on RBC deformability, biologic function and in vivo circulation. In evidence of this, both biological conditions and pharmacologic agents that affect hemoglobin content and/or viscosity or the RBC cytoskeleton alter cellular deformability and have profound in vivo and in vitro effects on RBC function and survival [5–16]. Indeed, RBC deformability can be a diagnostic indicator of RBC abnormalities and the quality of stored RBC prior to transfusion [17–28].

Intracellular viscosity and membrane structure are similarly key to the rheological deformability of WBC. However, in contrast to RBC, WBC intracellular viscosity is more complex and affected by multiple components including the: nuclear to cytoplasm (N:C) ratio; intracellular granule composition; presence of cytoplasmic vacuoles; as well as the activation state of the immune cell (**Figure 1B**) [28–30]. Similarly, membrane rigidity is also more complex due to: abundance of membrane proteins and protein rafts; changes in protein structure and polymerization consequent to immune activation; and the variability of the membrane and cytoskeletal protein composition of immune cell populations (e.g., monocytes, lymphocytes, granulocytes) and subsets (e.g., T cells versus B cells; CD4+ versus CD8+ T cells; NK cells) [30–35]. Perhaps surprisingly, despite the biologic importance of its rheological deformability within the vasculature, WBC deformability is both poorly defined and much less understood. Indeed, previous studies on WBC have most commonly defined “deformability” as cellular shape change or spreading under extrinsic suction (e.g., micropipette aspiration), compression pressure (e.g., centrifugation and cell poker/probe), or upon activation induced motility [30–32, 34–36]. However, vascular deformability is vastly different from cellular shape change or spreading which are most commonly induced by immune cell activation and, importantly, the actual loss of vascular rheologically-mediated (i.e., fluid motion and spatial confinement) deformability. The paucity of data relating to vascular deformability of WBC has, in large part, been due to the absence of suitable tools for measuring deformability across the broad range of cell types encompassed within leukocyte population. However, the complexity of the leukocyte population and resultant changes in rheological deformability upon activation (e.g., granule release) potentially arising in peripheral blood WBC may be of clinical importance as a biomarker of acute or chronic immune activation.

2. Measuring the vascular (rheological) deformability of blood cells

Because of the crucial role that cellular deformability plays in vascular circulation of RBC, methods to quantitate this biomechanical-aspect of normal and abnormal RBC has been of interest to hematologists since the 1960s [3, 5–7, 9, 10, 37–40]. Historically, multiple technological tools have been employed to study RBC (but rarely WBC) deformability including: micropipette aspiration; ektacytometry; cell transit times; and, most recently, microfluidic analysis.

2.1 Micropipette aspiration

Perhaps the earliest experimental approach to measure RBC deformability was the micropipette aspiration (**Figure 2**). Initial studies examined the ability of normal and stored RBC to traverse the length of a micropipette of known diameter [38]. This early "microfluidic" single cell analytical approach, while very low throughput and time consuming, did demonstrate that damaged or stored RBC were less deformable than fresh normal RBC. Subsequent variations of these micropipette studies further examined the localized elasticity of the membrane in both intact cells and RBC ghosts using ever smaller micropipettes to deform a small segment of the membrane to characterize static deformability via membrane extensional rigidity and bending rigidity. To further characterize dynamic deformability of the cells, the time constants for rapid elastic recovery from extensional and bending deformations were also quantitated [41–47]. However, micropipette, single-cell aspiration, measurements did not adequately reflect the biomechanical heterogeneity of even a relatively homogenous cell population (e.g., normal RBC), much less, the highly divergent population of cells encompassed within the WBC population. Hence newer methods were devised in an attempt to study large number of RBC under flow-like conditions. In contrast to RBC, micropipette studies are still commonly used to examine leukocytes; though these approaches tend not to be focused on rheological deformability [22, 35, 48–53].

2.2 Ektacytometry

Perhaps the most glaring flaw of the various micropipette aspiration approaches were their limitation to single cell analyses. To overcome this limitation, ektacytometry

Micropipette Aspiration	• Single cell analysis - low throughput • Microscopic analysis • Low cell numbers • Measures cortical tension • Membrane centric • Time-consuming • No sample recovery • Poor biological correlation

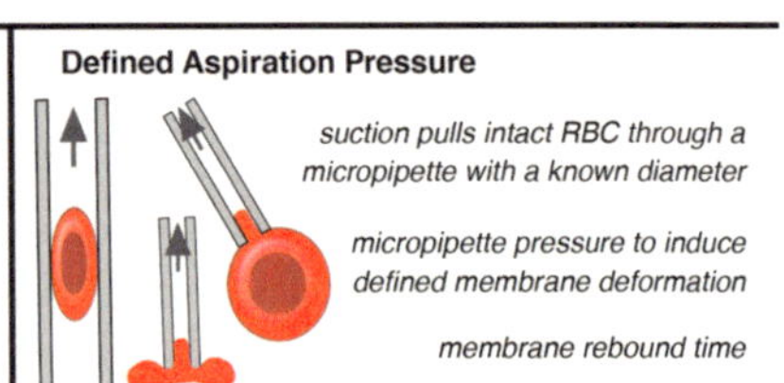

Figure 2.
Overview of micropipette aspiration analysis of blood cells. Micropipette-based analyses were first used to explore the crucial role of cellular deformability in the circulation of red blood cells. As noted, these single cell analyses were low throughput and time consuming. Multiple variation of this technique have been developed ranging from whole cell aspiration to localized membrane deformation. Studies could be done on intact cells or membrane ghosts.

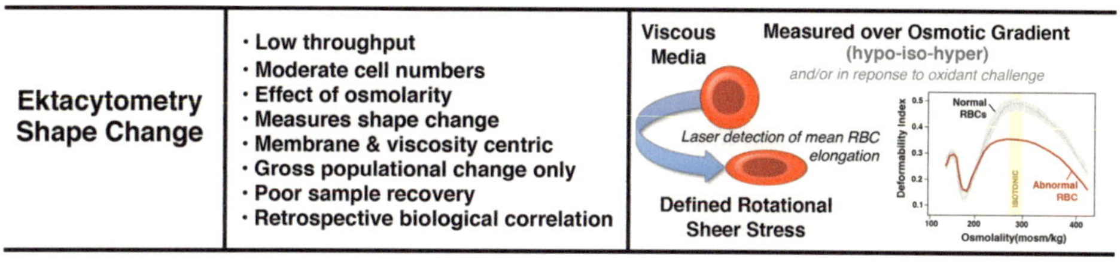

Figure 3.
Overview of ektacytometric analysis of blood cells. To overcome the single cell limitations inherent to micropipette aspiration, the ektacytometer can analyze the shear-induced deformability of a much larger population of RBC; though at the expense of information of single cell data acquisition. Ektacytometry measures deformability by subjecting RBC suspended in a viscous solution to rotational shear stress such that the normal cells form ellipsoids. The scatter intensity pattern from laser diffraction produces isointensity curves and deformability indices. Additionally, the most common approach of ektacytometry examines RBC deformability over a broad osmotic gradient (hypotonic → isotonic → hypertonic). Deformability is measured via laser diffraction as the shear stress forces the RBC to assume an elongated shape.

was developed. Ektacytometry measures deformability by suspending RBC in a viscous solution and applying rotational shear stress such that the normal discoid cells form ellipsoids which is measured by laser diffraction (**Figure 3**) [13, 14, 54–57]. The extent of ellipsoid formation is dependent on the deformability of the sample population. Abnormal RBC can be detected by shifts relative to the scatter intensity pattern of normal cells. Abnormal (i.e., non-deformable) cells can result in any combinations of left or right shifts in response to hypo- or hypertonicity, and/or a decrease in the maximum deformation observed under isotonic conditions. Relative to micropipette studies, ektacytometry provided a relative rapid assay to examine RBC. Numerous ektacytometry studies have elucidated the profound influence that mean corpuscular hemoglobin concentration (hence intracellular viscosity), abnormal hemoglobins, cytoskeletal aberrations, drugs and oxidant challenge exert on the cellular deformability [13, 14, 18, 54–64]. Importantly, ektacytometry only measures the "average deformability" of a cell population and cannot accurately and efficiently quantify the abundance of rigid cells in a bimodal population where both normal and abnormal cells are present [57, 65]. In the context of blood banking, ektacytometry has been used for assessing RBC following blood bank storage [66–68]. Of note, ektacytometry has been used exclusively in the context of erythrocytes; with no known studies examining the shear-induced deformability of lymphocytes, neutrophils, monocytes or other leukocytes. Thus, despite some promising data regarding its clinical use in transfusion medicine, ektacytometry has not become commonly used in transfusion medicine due to both the cost of instrumentation and the relatively low throughput of the existing testing protocols. Moreover, ektacytometry does have some significant drawbacks as it cannot, without experimental manipulations (e.g., density separation), provide any information on subsets of cells within the larger population—the results obtained are simply the "average" of the population. This limitation is, perhaps, the critical failure of ektacytometry because, in many pathologic states, abnormal RBC represent a minor (<10%) fraction of the overall RBC mass hence subtle changes will not be clearly obvious. Moreover, it is difficult to recover RBC subsequent to ektacytometric analysis for further biologic testing due to the viscous media utilized and, using traditional ektacytometry, the fact that the RBC are irreversibly (in most cases) altered by the osmotic gradient employed during the assay.

2.3 Cell transit analysis

In contrast to micropipette analysis and ektacytometry, cell transit analysis provides information at both the single cell and populational level (**Figure 4**).

Cell Transit Analysis	• Low throughput • Low cell numbers • Measures transit time (ms) • Membrane & viscosity centric • Gives population distribution • Poor sample recovery • Retrospective biological correlation • Early microfluidic approach	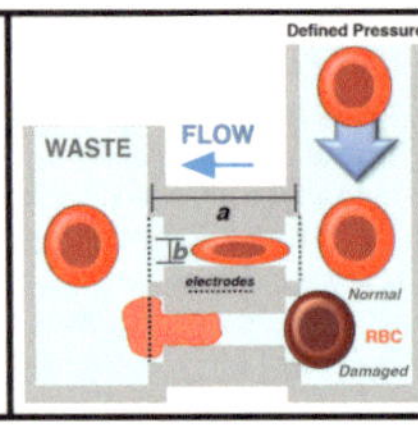	**Micropores of Defined Length (a) & Diameter (b)** **Computer Detected Impedence Transit Time (ms)** Damaged RBC occlude pores: rapidly degrade instrument

Figure 4.
Overview of cell transit analysis of blood cells. In contrast to micropipette analysis and ektacytometry, cell transit analysis provides information at both the single cell and populational level. Cellular deformability is indirectly measured via transit time (ms) of RBC through pores of defined diameter and length. Transit time is measured by the change in electrical resistance as an RBC passes through a micropore. Cell transit analysis is, in essence, an early micro (macro) fluidics approach.

To accomplish this, cell transit analysis combines features of both the traditional micropore filtration assay and the micropipette aspiration methodology, in that deformability of each RBC constitutes a single data point and can be used to then generate a populational distribution curve. In a cell transit analyzer, a single RBC passes through a micropore of fixed diameter and length with the transit time (in milliseconds; ms) of the cell calculated using the electrical resistance generated by the RBC within the channel as detected via a conductometer. However, the sensitivity of this method varies with cell size. Smaller cells, even if less deformable, pass through the pores with less resistance. In contrast, abnormally large or rigid cells, which are clinically important, are also be problematic as they block the micropore and are excluded from analysis [17, 69, 70]. Despite these limitations, cell transit analysis is very useful in that it provides subset/heterogeneity analysis via binning of the cells based on the transit time thus providing a continuous measure of the deformability profile of a sample and/or the severity of the deformability defect. The comparative utility of ektacytometry and cell transit analysis of RBC can be seen in normal and model ß thalassemic RBC in which purified alpha-hemoglobin chains are entrapped within normal RBC (**Figure 5**) [17, 19, 61–63]. While the ektacytometry and cell transit analysis have proven very useful as research tools, they have not been used to any great extent clinically. This is in large part due to the expense and complexity of the devices as well as their slow throughput making them impractical for clinical laboratories. Moreover, these in vitro studies often lack biological validation to the very low throughput of the assay (e.g., micropipette aspiration studies), overly small cell numbers, difficulty/impossibility of cell recovery post assessment, or more importantly, an inability to either identify or collect specific sample subsets (e.g., low versus high deformability) following analysis (e.g., Ektacytometry and Cell Transit Analysis studies).

2.4 Microfluidics

As noted in the preceding discussion, multiple micro/macro fluidic approaches have been used to model hemorheology of circulating blood cells; albeit almost exclusively RBC. Despite their valuable contributions to our understanding of blood cell deformability, these methods are inherently low throughput and dependent on relatively expensive instrumentation. But perhaps one of the biggest issues challenging these previous methodologies is the inability to recover substantial, or any, subpopulations (e.g., highly versus poorly deformable cells) from the analyzed sample. This weakness precludes additional in vitro or in vivo studies to tease out biological variations leading to the differential deformability profiles. Microfluidics approaches (**Figure 6**) potentially offers a cost-effective, high throughput,

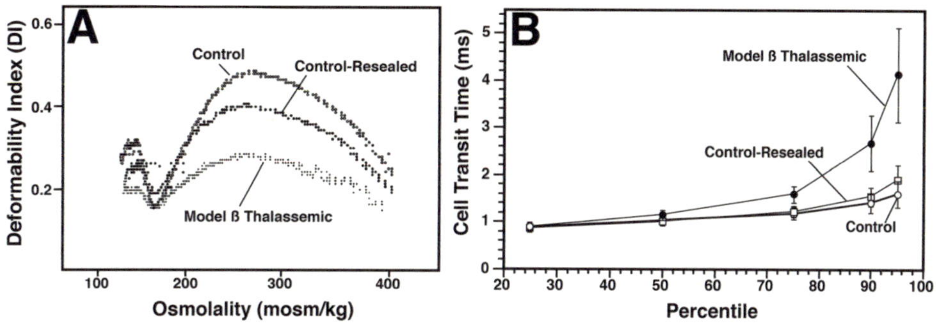

Figure 5.
Comparative data of normal versus model ß thalassemic RBC as assessed by ektacytometry and cell transit analysis. Panel A: as shown, ektacytometry provides the mean diffraction profile of a population of cells over a broad osmotic gradient; however, it does not provide any information as to the deformability distribution of the cells within the total population. Panel B: cell transit analysis gives information regarding both the deformability profile of the entire population and the individual cells within the tested population. Data derived from Refs. [17, 63].

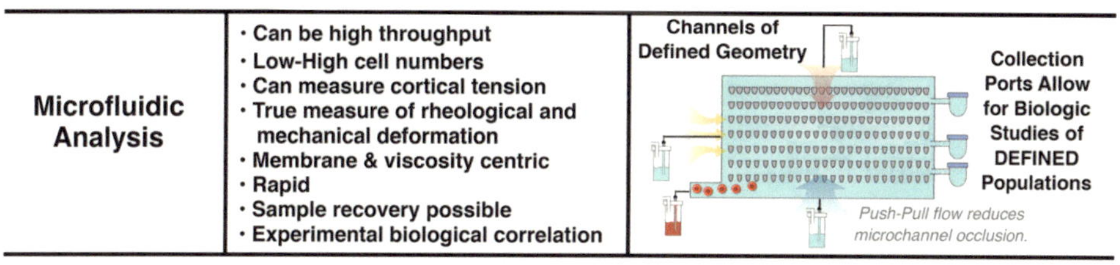

Figure 6.
Overview of microfluidic analysis of blood cells. Recent advances in microscale fabrication technologies have allowed for the development of an exceedingly broad array of microfluidic devices that may have utility in assessing the deformability of blood cells. These approaches range from single to multi-channel devices with channels of single or variable lengths and diameters. In addition, some designs incorporate collection ports so that cell exhibiting differential deformability profiles can be collected for further in vitro or in vivo study. Device shown is adapted from Guo et al. and Kang et al. [26, 28].

alternative to assessing blood cell deformability relative to these previous, and now rather ancient (as reflected by the key research papers relating to these approaches) technologies [22–25, 27, 28, 71–78]. Deformability measurement using microfluidics uses minute amounts of a whole blood or purified RBC/WBC in suspension flowing through a funnel-shaped micro-constriction(s) in a disposable plate. As demonstrated in our previous publications, and discussed in the following section, microfluidics devices are capable of providing reproducible intra- and inter-individual data, detecting oxidatively damaged RBC, identifying changes in RBC deformability consequent to storage, and identifying leukocytes [20–28].

3. Utility of microfluidics in transfusion medicine

As evidenced by the number of publications and patents being generated annually, the promise of microfluidic devices in medicine is seemingly unbounded. One area of particular interest to our laboratories has been in the field of transfusion medicine [20–28]. Annually over 100,000,000 units of blood are collected worldwide for transfusion purposes. Despite the volume collected, our tools for assessing the quality of the stored blood products remains primarily centered on 1950–80s technology. Upon collection of whole blood in Canada the blood is processed to produce 3 major components: RBC, platelets and plasma. The RBC component for use in blood transfusion therapy are stored at 4°C for up to 42 days. The maximum storage window for RBC is based on studies dating from the 1950s on that defined a ≥ 75% recovery rate at 24 hours post-transfusion as the clinical "quality control" standard for stored donor RBC [79, 80]. Despite decades of research into RBC biology and advances in other aspects of transfusion medicine, the 24 hour survival rule remains the current gold standard for determining acceptable donor RBC quality in transfusion medicine. Currently there are no other established biomarkers by which blood services can discriminate "good" versus "bad" units. Note however, that ultimately the survival of the donor RBC is consequent to their vascular deformability (which is in turn governed by a multitude of biologic/metabolic factors). Hence, cost effectively assessing the deformability of stored RBC could serve as an excellent biomarker for the quality of stored donor RBC. Intriguingly, RBC deformability may also be a potent pre-screening tool that could be used to exclude potential donors from RBC donations. RBC which demonstrate poor initial deformability upon collection do not store well and may lead to adverse events in patients who receive these units. Poor deformability of potential donor RBC may arise from a broad range of issues including: undiagnosed RBC abnormalities (e.g., cytoskeletal, hemoglobin or metabolic aberrations); vascular inflammation; or dietary or drug-mediated alterations of the RBC.

To assess the deformability of blood cells, our laboratories have utilized a variety of microfluidic devices ranging from a simple, low throughput, funnel chain (prone to clogging) to a much more advanced and robust high throughput ratchet device. The ratchet microfluidic approach has proved better at assessing vascular deformability as blood cells are pushed laterally and vertically through tapered microchannels of decreasing size thus modeling the process of cellular deformation in microvasculature (**Figure** 7). Vertical movement is done via an oscillatory vertical pressure deferential that allows both a net vertical filtration flow and a downward declogging flow to minimize microchannel obstruction by blood cells as they reach their deformability limit. Importantly, this design also incorporates collection outlets allowing for recovery, and further testing, of cell populations with differential deformability profiles. Our research to date has demonstrated that this microfluidic microfiltration device is capable of isolating circulating tumor cells

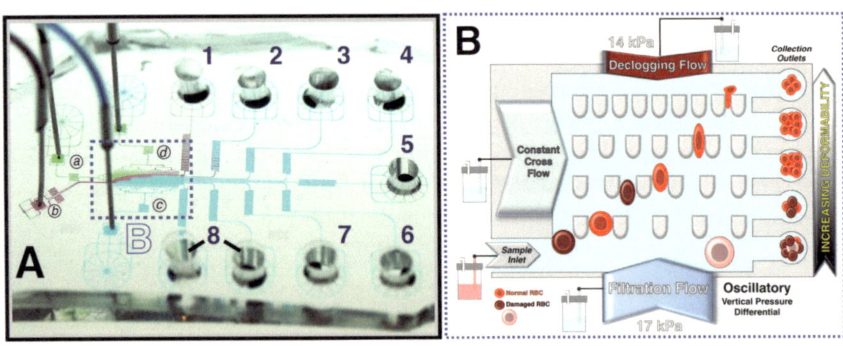

Figure 7.
General schematic of a ratchet microfluidic device. Panel A: shown is a photograph of the ratchet microfluidic device infused with different color dyes to highlight the design features: cross flow inlet (a), sample inlet (b), upward (c) and downward (d) oscillatory flow inlets, sorting region (dashed blue box) and outlets 1–8. In this design, outlets 8–1 corresponds to blocking pore sizes of ≥6.5, 5.5, 4.5, 4.0, 3.5, 3.0, 2.5 and 2.0 μm, respectively. Panel B: schematic of the sorting region showing the decreasing size of the tapered microchannels as well as the deformability of normal and oxidized RBC through these microchannels. Poorly deformable cells (e.g., oxidized RBC) are collected in outlets 8 and 7 while highly deformable cells are collected in outlets 3 thru 1. The downward oscillatory pressure minimizes channel obstruction by poorly deformable cells which are pushed horizontally into the collection outlets by the cross flow pressure. This device is suitable for use for both human RBC and WBC.

from leukocytes, malaria-infected and oxidized RBC from normal cells, granulocytes and lymphocytes from whole blood, and detecting early immune cell activation consequent to degranulation [26–28, 81].

Key to the use of microfluidic devices in RBC blood banking is documenting the ability of the device(s) to discriminate between "normal" and abnormal cellular deformability and document that the loss of deformability is associated with diminished in vivo circulation. Loss of cellular deformability can arise from a host of causes, most of which, due to the iron and oxygen rich environment of the RBC, leads to cellular oxidation [17–19, 23, 57, 61, 63, 82]. As shown in **Figure 8**, human or murine RBC oxidized by exposure to 50 μM phenazine methosulfate (PMS) were readily discriminated from normal RBC as measured by the cortical tension required to push the RBC through a funnel shaped micropore However, as noted by the differences between the human and murine RBC, the microchannel size (2–2.5 μm in this experiment) relative to the mean diameter of the RBC itself (~8 versus 6.7 μm for human and mouse RBC, respectively) will also play a role. Most importantly however, the loss of murine deformability in the oxidized RBC sample, as noted in the microfluidic device, correlated closely with the loss of in vivo survival. These findings suggest that microfluidic devices could prove useful for both diagnostic purposes (e.g., hemoglobinopathies such as sickle cell disease and thalassemia) as well as in evaluating the quality of stored human RBC prior to transfusion into a patient.

Indeed, microfluidics analysis of stored human RBC suggests that deformability is affected by storage time. As demonstrated by Matthews et al., using a microfluidic device, there is a significant loss of RBC deformability as early as 2 weeks into storage [25]. This finding confirms single-cell deformability studies that similarly

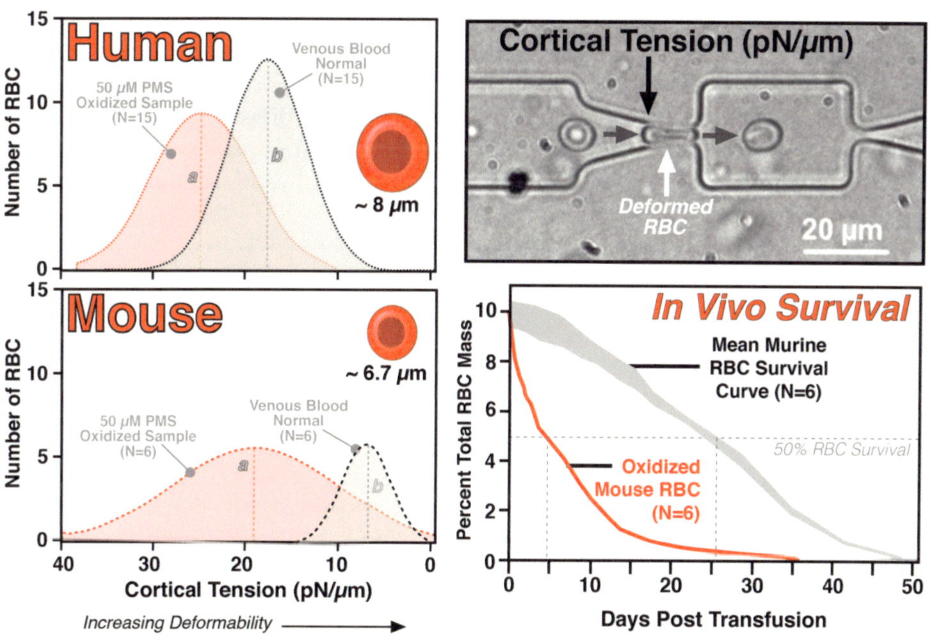

Figure 8.
Analysis of human and murine RBC deformability using a conical microfluidic array. The width of the funnel shaped micropore constriction used to measure RBC deformability was approximately 2–2.5 μm in size at its minimum. (a) and (b) equals peak count for oxidized and normal RBC respectively. Human blood was obtained via a finger prick while mouse blood was obtained by saphenous bleed. Also noted is the 50% in vivo survival point for oxidized (~5 days) and normal (~26 days) murine RBC. Data derived from Kwan et al. [23].

indicated that RBC deformability remained fairly constant in the first 2–3 weeks of storage and then rapidly decreased [83, 84]. However, in contrast to these single cell studies, our high throughput device can rapidly assess the proportion of individual RBCs that are too rigid to transit the microconstrictions and may, upon transfusion into an individual, be cleared by the spleen. Indeed, by day 42 of storage, 30% of all donor RBCs were too rigid to transit the device. Interestingly, a small subset of donors had RBC that demonstrated poor storage in that >50% of their RBC were too rigid to passage the microconstriction. These research findings suggest that the RBC quality of individual donors are, not unexpectedly, variable. The source of inter-individual variability causing the poor storage could be either inherent to the donor RBC itself (e.g., metabolic, structural or hemoglobin abnormalities) or transient (e.g., inflammation, food or drug induced).

The prescreening questionnaire completed by both new and repeat blood donors is focused, in part, on identifying factors that could adversely affect the quality of the blood product(s) produced from a donation. While most biologically-mediated RBC defects are likely to have been previously detected during normal medical surveillance of the prospective donor, transient inflammatory-mediated effects, such as those arising from viral, bacterial, drug or autoimmune events, are most likely to impact blood component quality. To address these potential risks, at the time of blood donation, all donors are asked if they feel ill or have had a recent fever. While the primary purpose of these self-reporting questions is to avoid transfusion of blood-borne infective agents or plasma that may contain potent immunomodulatory chemokines and cytokines, systemic inflammatory events may also result in bystander injury to the RBC that may compromise RBC storage and safety. The described microfluidics ratchet device may also provide a means of assessing both the WBC population and activation state of an individual [26, 28]. As shown in **Figure 9**, the ratchet microfluidic device described in **Figure 7**, is capable of differentially sorting monocytes from lymphocytes. The same device can also differentiate between resting (granule containing) from activated (degranulated) CD8+ T lymphocytes. Further refinement of the microchannel geometry will be capable of improving cell separation making it possible to readily prescreen individuals for evidence of immune activation thus improving blood component safety consequent to empirical donor evaluation versus self-reporting. Finally, microfluidic devices

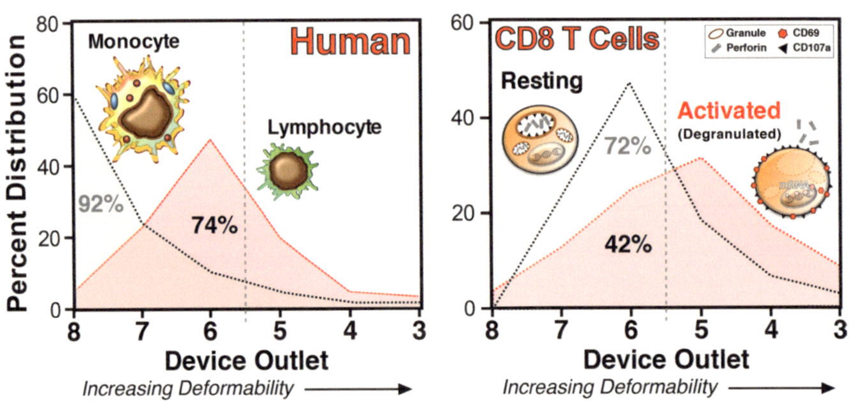

Figure 9.
Analysis of human monocyte and lymphocyte populations showing differential sorting on the ratchet microfluidic device. Note that the prototype device can also detect degranulation of lymphocytes (shown are CD8+ T cells) which occurs upon inflammatory activation. The vertical dashed line separates cells based on less deformable (collection outlets 8–6) and more deformable (collection outlets 5–3). Data derived from Kang et al. [28].

could also be used during the blood collection process, as well as in the field, to screen individuals who have reported recent travel to malarial endemic areas, for actual malaria infection [27, 85–88]. Currently, individuals traveling to malarial endemic regions are deferred from blood donation; an action that often results in their permanent loss from the blood donor pool.

4. Conclusions

Microfluidics devices have the potential to dramatically, and cost-effectively, change the practice of transfusion medicine. As illustrated, purpose-specific development of ratchet microfluidics devices will make it possible, via a finger prick (e.g., as shown in **Figure 8**), to prescreen donors at the time of pre-donation testing (i.e., simultaneously with determining the donor's hematocrit prior to unit donation) to select donors whose RBC show normal deformability profiles prior to storage. Donors with RBC deformability profiles outside of the normal range would be deferred from RBC donation, though potentially, still donating plasma for fractionation into plasma protein components. Moreover, the same microfluidic approach could improve the detection of patients with recent/current systemic immune activation that could result in the presence of undesirable cytokines/chemokines within the donated blood or that might have adversely affected normal RBC deformability. Hence, the cost-effective microfluidic-based prescreening process would potentially diminish the risk to patient safety that accompanies ineffectual RBC transfusion and/or the presence of inflammatory mediators in blood products. Not inconsequentially, prescreening for good donors would reduce the expense to the blood operator associated with the production and distribution of a potentially ineffectual, or unsafe, blood unit. Beyond prescreening donors, patient safety would also be enhanced by doing point-of-care deformability analysis of stored RBC prior to transfusion. Such analysis would enhance patient safety by reducing the aggregate transfusion needs of a patient by preventing the transfusion of RBC which would have poor in vivo survivability. Such an approach would be of particular value in the chronically transfused patient (e.g., sickle cell, thalassemic and myelodysplastic) populations.

Acknowledgements

This work was supported by grants from the Canadian Institutes of Health Research (325,373, HM and MDS, 322375, HM; and 362,500, HM), Canadian Blood Services-CIHR Partnership program (BUC21403-HM; HM and MDS), Canadian Blood Services (MDS) and Health Canada (MDS). The views expressed herein do not necessarily represent the view of the federal government of Canada. We thank the Canada Foundation for Innovation and the Michael Smith Foundation for Health Research for infrastructure funding at the University of British Columbia Centre for Blood Research. The funders had no role in study design, data collection and analysis, decision to publish, or preparation of the manuscript.

Conflict of interest

The University of British Columbia and HM have pending patent applications relating to the described microfluidic devices.

Author details

Mark D. Scott[1,2,3*], Kerryn Matthews[2,4] and Hongshen Ma[2,4]

1 Centre for Innovation, Canadian Blood Services, University of British Columbia, Vancouver, BC, Canada

2 Centre for Blood Research, University of British Columbia, Vancouver, BC, Canada

3 Department of Pathology and Laboratory Medicine, University of British Columbia, Vancouver, BC, Canada

4 Department of Mechanical Engineering, University of British Columbia, Vancouver, BC, Canada

*Address all correspondence to: mdscott@mail.ubc.ca

References

[1] Aird WC. Spatial and temporal dynamics of the endothelium. Journal of Thrombosis and Haemostasis. 2005;**3**:1392-1406

[2] Wexler L, Bergel DH, Gabe IT, Makin GS, Mills CJ. Velocity of blood flow in normal human venae cavae. Circulation Research. 1968;**23**:349-359

[3] Weiss L, Tavassoli M. Anatomical hazards to the passage of erythrocytes through the spleen. Seminars in Hematology. 1970;**7**:372-380

[4] Chen LT, Weiss L. The role of the sinus wall in the passage of erythrocytes through the spleen. Blood. 1973;**41**:529-537

[5] LaCelle PL. Alteration of membrane deformability in hemolytic anemias. Seminars in Hematology. 1970;**7**:355-371

[6] Weed RI. The importance of erythrocyte deformability. The American Journal of Medicine. 1970;**49**:147-150

[7] Chien S, Usami S, Bertles JF. Abnormal rheology of oxygenated blood in sickle cell anemia. The Journal of Clinical Investigation. 1970;**49**:623-634

[8] Chien S, Usami S, Dellenback RJ, Gregersen MI. Shear-dependent deformation of erythrocytes in rheology of human blood. The American Journal of Physiology. 1970;**219**:136-142

[9] Bessis M, Mohandas N. Red cell structure, shapes and deformability. British Journal of Haematology. 1975;**31**:5-11

[10] La Celle PL. Pathogenic erythrocytes in the capillary microcirculation. Blood Cells. 1975;**1**:269-284

[11] Havell TC, Hillman D, Lessin LS. Deformability characteristics of sickle cells by microelastimetry. American Journal of Hematology. 1978;**4**:9-16

[12] Mohandas N, Phillips WM, Bessis M. Red blood cell deformability and hemolytic anemias. Seminars in Hematology. 1979;**16**:95-114

[13] Clark MR, Mohandas N, Shohet SB. Deformability of oxygenated irreversibly sickled cells. The Journal of Clinical Investigation. 1980;**65**:189-195

[14] Clark MR, Mohandas N, Shohet SB. Osmotic gradient ektacytometry: Comprehensive characterization of red cell volume and surface maintenance. Blood. 1983;**61**:899-910

[15] Snyder LM, Fortier NL, Trainor J, Jacobs J, Leb L, Lubin B, et al. Effect of hydrogen peroxide exposure on normal human erythrocyte deformability, morphology, surface characteristics, and spectrin-hemoglobin cross-linking. The Journal of Clinical Investigation. 1985;**76**:1971-1977

[16] Snyder LM, Fortier NL, Leb L, McKenney J, Trainor J, Sheerin H, et al. The role of membrane protein sulfhydryl groups in hydrogen peroxide-mediated membrane damage in human erythrocytes. Biochimica et Biophysica Acta. 1988;**937**:229-240

[17] Scott MD, Rouyer-Fessard P, Ba MS, Lubin BH, Beuzard Y. Alpha- and beta-haemoglobin chain induced changes in normal erythrocyte deformability: Comparison to beta thalassaemia intermedia and Hb H disease. British Journal of Haematology. 1992;**80**:519-526

[18] Scott MD, van den Berg JJ, Repka T, Rouyer-Fessard P, Hebbel RP, Beuzard Y, et al. Effect of excess alpha-hemoglobin chains on cellular and membrane oxidation in model beta-thalassemic erythrocytes. The Journal of Clinical Investigation. 1993;**91**:1706-1712

[19] Scott MD. H2O2 injury in beta thalassemic erythrocytes: Protective role of catalase and the prooxidant effects of GSH. Free Radical Biology & Medicine. 2006;**40**:1264-1272

[20] Guo Q, McFaul SM, Ma H. Deterministic microfluidic ratchet based on the deformation of individual cells. Physical Review E, Statistical, Nonlinear, and Soft Matter Physics. 2011;**83**:051910

[21] Guo Q, Reiling SJ, Rohrbach P, Ma H. Microfluidic biomechanical assay for red blood cells parasitized by plasmodium falciparum. Lab on a Chip. 2012;**12**:1143-1150

[22] Guo Q, Park S, Ma H. Microfluidic micropipette aspiration for measuring the deformability of single cells. Lab on a Chip. 2012;**12**:2687-2695

[23] Kwan JM, Guo Q, Kyluik-Price DL, Ma H, Scott MD. Microfluidic analysis of cellular deformability of normal and oxidatively damaged red blood cells. American Journal of Hematology. 2013;**88**:682-689

[24] Guo Q, Duffy SP, Matthews K, Santoso AT, Scott MD, Ma H. Microfluidic analysis of red blood cell deformability. Journal of Biomechanics. 2014;**47**:1767-1776

[25] Matthews K, Myrand-Lapierre ME, Ang RR, Duffy SP, Scott MD, Ma H. Microfluidic deformability analysis of the red cell storage lesion. Journal of Biomechanics. 2015;**48**:4065-4072

[26] Guo Q, Duffy SP, Matthews K, Islamzada E, Ma H. Deformability based cell sorting using microfluidic ratchets enabling phenotypic separation of leukocytes directly from whole blood. Scientific Reports. 2017;**7**:6627

[27] Matthews K, Duffy SP, Myrand-Lapierre ME, Ang RR, Li L, Scott MD, et al. Microfluidic analysis of red blood cell deformability as a means to assess hemin-induced oxidative stress resulting from plasmodium falciparum intraerythrocytic parasitism. Integrative Biology. 2017;**9**:519-528

[28] Kang N, Guo Q, Islamzada E, Ma H, Scott MD. Microfluidic determination of lymphocyte vascular deformability: Effects of intracellular complexity and early immune activation. Integrative Biology. 2018;**10**:207-217

[29] Abbas AK, Lichtman AH, Pillai S. Cellular and Molecular Immunology. Philadelphia, Pennsylvania, USA: Elsevier/Saunders; 2014:13-34

[30] Rosenbluth MJ, Lam WA, Fletcher DA. Force microscopy of nonadherent cells: A comparison of leukemia cell deformability. Biophysical Journal. 2006;**90**:2994-3003

[31] Mege JL, Capo C, Benoliel AM, Foa C, Bongrand P. Study of cell deformability by a simple method. Journal of Immunological Methods. 1985;**82**:3-15

[32] Pasternak C, Elson EL. Lymphocyte mechanical response triggered by cross-linking surface receptors. The Journal of Cell Biology. 1985;**100**:860-872

[33] Downey GP, Doherty DE, Schwab B, Elson EL, Henson PM, Worthen GS. Retention of leukocytes in capillaries: Role of cell size and deformability. Journal of Applied Physiology. 1990;**69**:1767-1778

[34] Brown MJ, Hallam JA, Colucci-Guyon E, Shaw S. Rigidity of circulating lymphocytes is primarily conferred by vimentin intermediate filaments. Journal of Immunology. 2001;**166**:6640-6646

[35] Esteban-Manzanares G, González-Bermúdez B, Cruces J, De la Fuente M, Li Q, Guinea GV, et al.

Improved measurement of elastic properties of cells by micropipette aspiration and its application to lymphocytes. Annals of Biomedical Engineering. 2017;**45**:1375-1385

[36] Zhang X, Cook PC, Zindy E, Williams CJ, Jowitt TA, Streuli CH, et al. Integrin α4β1 controls G9a activity that regulates epigenetic changes and nuclear properties required for lymphocyte migration. Nucleic Acids Research. 2016;**44**:3031-3044

[37] Nevaril CG, Lynch EC, Alfrey CP, Hellums JD. Erythrocyte damage and destruction induced by shearing stress. The Journal of Laboratory and Clinical Medicine. 1968;**71**:784-790

[38] La Celle PL. Alteration of deformability of the erythrocyte membrane in stored blood. Transfusion. 1969;**9**:238-245

[39] Weed RI, LaCelle PL, Merrill ET. Erythrocyte metabolism and cellular deformability. Vox Sanguinis. 1969;**17**:32-33

[40] Weed RI, LaCelle PL, Merrill EW. Metabolic dependence of red cell deformability. The Journal of Clinical Investigation. 1969;**48**:795-809

[41] Evans E, Mohandas N, Leung A. Static and dynamic rigidities of normal and sickle erythrocytes. Major influence of cell hemoglobin concentration. The Journal of Clinical Investigation. 1984;**73**:477-488

[42] Evans EA, Mohandas N. Membrane-associated sickle hemoglobin: A major determinant of sickle erythrocyte rigidity. Blood. 1987;**70**:1443-1449

[43] Ballas SK, Larner J, Smith ED, Surrey S, Schwartz E, Rappaport EF. Rheologic predictors of the severity of the painful sickle cell crisis. Blood. 1988;**72**:1216-1223

[44] Mohandas N, Evans E. Mechanical properties of the red cell membrane in relation to molecular structure and genetic defects. Annual Review of Biophysics and Biomolecular Structure. 1994;**23**:787-818

[45] Discher DE, Mohandas N, Evans EA. Molecular maps of red cell deformation: Hidden elasticity and in situ connectivity. Science. 1994;**266**:1032-1035

[46] Heinrich V, Ritchie K, Mohandas N, Evans E. Elastic thickness compressibilty of the red cell membrane. Biophysical Journal. 2001;**81**:1452-1463

[47] Evans J, Gratzer W, Mohandas N, Parker K, Sleep J. Fluctuations of the red blood cell membrane: Relation to mechanical properties and lack of ATP dependence. Biophysical Journal. 2008;**94**:4134-4144

[48] Schmid-Schönbein GW, Sung KL, Tözeren H, Skalak R, Chien S. Passive mechanical properties of human leukocytes. Biophysical Journal. 1981;**36**:243-256

[49] Derganc J, Bozic B, Svetina S, Zeks B. Stability analysis of micropipette aspiration of neutrophils. Biophysical Journal. 2000;**79**:153-162

[50] Shao JY, Xu J. A modified micropipette aspiration technique and its application to tether formation from human neutrophils. Journal of Biomechanical Engineering. 2002;**124**:388-396

[51] Liu B, Goergen CJ, Shao JY. Effect of temperature on tether extraction, surface protrusion, and cortical tension of human neutrophils. Biophysical Journal. 2007;**93**:2923-2933

[52] Kaleridis V, Athanassiou G, Deligianni D, Missirlis Y. Slow flow of passive neutrophils and sequestered nucleus into micropipette. Clinical

Hemorheology and Microcirculation. 2010;**45**:53-65

[53] Guillou L, Babataheri A, Saitakis M, Bohineust A, Dogniaux S, Hivroz C, et al. T-lymphocyte passive deformation is controlled by unfolding of membrane surface reservoirs. Molecular Biology of the Cell. 2016;**27**:3574-3582

[54] Kuypers FA, Chiu D-Y, Mohandas N, Roelofsen B, Op den Kamp JAF, Lubin BH. The molecular species composition of phosphatidylcholine affects cellular properties in normal and sickle erythrocytes. Blood. 1987;**70**:1111-1118

[55] Green MA, Noguchi CT, Keidan AJ, Marwah SS, Stuart J. Polymerization of sickle cell hemoglobin at arterial oxygen saturation impairs erthrocyte deformability. The Journal of Clinical Investigation. 1988;**81**:1669-1674

[56] Chasis JA, Schrier SL. Membrane deformability and the capacity for shape change in the erythrocyte. Blood. 1989;**74**:2562-2568

[57] Kuypers FA, Scott MD, Schott MA, Lubin B, Chiu DT. Use of ektacytometry to determine red cell susceptibility to oxidative stress. The Journal of Laboratory and Clinical Medicine. 1990;**116**:535-545

[58] Scott MD, Meshnick SR, Williams RA, Chiu D-Y, Lubin FA, Kuypers FA. Qinghaosu-enhanced oxidant sensitivity in erythrocytes with unstable hemoglobins. Blood. 1988;**72**:200

[59] Scott MD, Eaton JW, Kuypers FA, Chiu D-Y, Lubin BH. Enhancement of erythrocyte superoxide dismutase activity: Effects on cellular oxidant defense. Blood. 1989;**74**:2542-2549

[60] Butikofer P, Lin ZW, Kuypers FA, Scott MD, Xu CM, Wagner GM, et al. Chlorpromazine inhibits vesiculation, alters phosphoinositide turnover and changes deformability of ATP-depleted RBCs. Blood. 1989;**73**:1699-1704

[61] Scott MD, Rouyer-Fessard P, Lubin BH, Beuzard Y. Entrapment of purified alpha-hemoglobin chains in normal erythrocytes. A model for beta thalassemia. The Journal of Biological Chemistry. 1990;**265**:17953-17959

[62] Scott MD, Kuypers FA, Butikofer P, Bookchin RM, Ortiz OE, Lubin BH. Effect of osmotic lysis and resealing on red cell structure and function. The Journal of Laboratory and Clinical Medicine. 1990;**115**:470-480

[63] Kuypers FA, Schott MA, Scott MD. Phospholipid composition and organization in model beta-thalassemic erythrocytes. American Journal of Hematology. 1996;**51**:45-54

[64] Murad KL, Mahany KL, Brugnara C, Kuypers FA, Eaton JW, Scott MD. Structural and functional consequences of antigenic modulation of red blood cells with methoxypoly(ethylene glycol). Blood. 1999;**93**:2121-2127

[65] Streekstra GJ, Dobbe JG, Hoekstra AG. Quantification of the fraction poorly deformable red blood cells using ektacytometry. Optics Express. 2010;**18**:14173-14182

[66] Frank SM, Abazyan B, Ono M, Hogue CW, Cohen DB, Berkowitz DE, et al. Decreased erythrocyte deformability after transfusion and the effects of erythrocyte storage duration. Anesthesia and Analgesia. 2013;**116**:975-981

[67] Reinhart WH, Piety NZ, Deuel JW, Makhro A, Schulzki T, Bogdanov N, et al. Washing stored red blood cells in an albumin solution improves their morphologic and hemorheologic properties. Transfusion. 2015;**55**:1872-1881

[68] Nagababu E, Scott AV, Johnson DJ, Dwyer IM, Lipsitz JA, Barodka VM, et al. Oxidative stress and rheologic properties of stored red blood cells before and after transfusion to surgical patients. Transfusion. 2016;**56**:1101-1111

[69] Baskurt OK. Deformability of red blood cells from different species studied by resistive pulse shape analysis technique. Biorheology. 1996;**33**:169-179

[70] OK B, TC F, HJ M. Sensitivity of the cell transit analyzer (CTA) to alterations of red blood cell deformability: Role of cell size-pore size ratio and sample preparation. Clinical Hemorheology. 1996;**16**:753-765

[71] Whitesides GM. The origins and the future of microfluidics. Nature. 2006;**442**:368-373

[72] Shevkoplyas SS, Yoshida T, Gifford SC, Bitensky MW. Direct measurement of the impact of impaired erythrocyte deformability on microvascular network perfusion in a microfluidic device. Lab on a Chip. 2006;**6**:914-920

[73] Xia N, Hunt TP, Mayers BT, Alsberg E, Whitesides GM, Westervelt RM, et al. Combined microfluidic-micromagnetic separation of living cells in continuous flow. Biomedical Microdevices. 2006;**8**:299-308

[74] Bransky A, Korin N, Nemirovski Y, Dinnar U. Correlation between erythrocytes deformability and size: A study using a microchannel based cell analyzer. Microvascular Research. 2007;**73**:7-13

[75] Forsyth AM, Wan J, Ristenpart WD, Stone HA. The dynamic behavior of chemically "stiffened" red blood cells in microchannel flows. Microvascular Research. 2010;**80**:37-43

[76] Ye T, Li H, Lam KY. Modeling and simulation of microfluid effects on deformation behavior of a red blood cell in a capillary. Microvascular Research. 2010;**80**:453-463

[77] Martin JD, Marhefka JN, Migler KB, Hudson SD. Interfacial rheology through microfluidics. Advanced Materials. 2011;**23**:426-432

[78] Patel KV, Mohanty JG, Kanapuru B, Hesdorffer C, Ershler WB, Rifkind JM. Association of the red cell distribution width with red blood cell deformability. Advances in Experimental Medicine and Biology. 2013;**765**:211-216

[79] Bratosin D, Estaquier J, Ameisen JC, Montreuil J. Molecular and cellular mechanisms of erythrocyte programmed cell death: Impact on blood transfusion. Vox Sanguinis. 2002;**83**(Suppl. 1):307-310

[80] Dumont LJ, AuBuchon JP. Evaluation of proposed FDA criteria for the evaluation of radiolabeled red cell recovery trials. Transfusion. 2008;**48**:1053-1060

[81] Park ES, Jin C, Guo Q, Ang RR, Duffy SP, Matthews K, et al. Continuous flow deformability-based separation of circulating tumor cells using microfluidic ratchets. Small. 2016;**12**:1909-1919

[82] Scott MD, Eaton JW. Thalassaemic erythrocytes: Cellular suicide arising from iron and glutathione-dependent oxidation reactions. British Journal of Haematology. 1995;**91**:811-819

[83] Czerwinska J, Rieger M, Uehlinger DE. Dynamics of red blood cells in microporous membranes. Biomicrofluidics. 2014;**8**:044101

[84] Huang S, Hou HW, Kanias T, Sertorio JT, Chen H, Sinchar D, et al. Towards microfluidic-based depletion of stiff and fragile human red cells that

accumulate during blood storage. Lab on a Chip. 2015;**15**:448-458

[85] Santoso AT, Deng X, Lee JH, Matthews K, Duffy SP, Islamzada E, et al. Microfluidic cell-phoresis enabling high-throughput analysis of red blood cell deformability and biophysical screening of antimalarial drugs. Lab on a Chip. 2015;**15**:4451-4460

[86] Myrand-Lapierre ME, Deng X, Ang RR, Matthews K, Santoso AT, Ma H. Multiplexed fluidic plunger mechanism for the measurement of red blood cell deformability. Lab on a Chip. 2015;**15**:159-167

[87] Deng X, Duffy SP, Myrand-Lapierre ME, Matthews K, Santoso AT, Du YL, et al. Reduced deformability of parasitized red blood cells as a biomarker for anti-malarial drug efficacy. Malaria Journal. 2015;**14**:428

[88] Guo Q, Duffy SP, Matthews K, Deng X, Santoso AT, Islamzada E, et al. Deformability based sorting of red blood cells improves diagnostic sensitivity for malaria caused by plasmodium falciparum. Lab on a Chip. 2016;**16**:645-654